Biochemistry

Biochemistry

An Illustrated Review with Questions and Explanations

Fifth Edition

Paul Jay Friedman, M.D.
Director of Rehabilitation, Waikato Hospital,
Clinical Division of Auckland University School of Medicine,
Hamilton, New Zealand

Little, Brown and Company
Boston   New York   Toronto   London

Library of Congress Cataloging-in-Publication Data
Friedman, Paul Jay.
    Biochemistry : a review with questions and explanations / Paul Jay
Friedman. — 5th ed.
        p.   cm.
    Includes bibliographical references and index.
    ISBN 0-316-29428-4 (pbk.)
    1. Biochemistry.   2. Biochemistry—Examinations, questions, etc.
I. Title.
    [DNLM: 1. Biochemistry—examination questions.   QU 18 F911b 1994]
QP514.2.F743   1994
574.19′2′076—dc20
DNLM/DLC
for Library of Congress   94-14756
                                                                                 CIP

Printed in the United States of America
SEM

Editorial: Evan R. Schnittman
Production Editor: Marie A. Salter
Copyeditor: Sharon Cloud Hogan
Production Supervisor/Designer: Michael A. Granger
Cover Designer: Michael A. Granger
Cover Illustration: Natalie C. Johnson

To Ralph B. Friedman, my father and the best teacher I've ever had, and Diane A. Friedman, my wife, who provided the encouragement and assistance I needed to complete this book

Contents

Preface

All the atoms of the earth bear witness, O my Lord, to the greatness of Thy power and of Thy sovereignty; and all the signs of the universe attest to the glory of Thy might.

GLEANINGS FROM THE WRITINGS OF BAHA'U'LLAH

This medical review text was written to explain the fundamentals of biochemistry to medical students and to cover the major topics in a medical biochemistry course and on the USMLE (Step I). I suggest that you use this book in concert with one or more of the standard biochemistry texts cited in the references at the end of each chapter. The illustrations found in the larger texts will assist you in understanding the material in this book.

When I was a medical student at the University of Illinois, I found that most existing biochemistry texts were more than 1000 pages in length. Much of the information they contained was unrelated to medical biochemistry. Furthermore, topics vital to human biochemistry were often covered insufficiently.

My aim in writing *Biochemistry* was to make learning as it should be—enjoyable. Each chapter in this book is short and easy to digest. The content of each focuses on the essential information. Because people learn best when they ask questions and then seek appropriate answers, each chapter comes equipped with a set of questions along with their step-by-step solutions. Through sharpening their problem-solving skills, students develop a comprehension of biochemistry that reduces the need for thoughtless memorization on the night before an exam. In preparing the fifth edition I have revised most of the questions, putting them into the format usually seen on the USMLE (Step I). Some questions, however, have been left open-ended because there is value to be gained in solving them without the constraints of multiple-choice or matching responses.

The fifth edition of *Biochemistry* represents a major change from the previous edition. I have changed the order of chapters to facilitate learning. Thus, discussion of carbohydrate structure in Chapter 7 is followed by discussion of carbohydrate metabolism in Chapter 8. Likewise, nucleic acid and lipid metabolism are covered in the chapters that immediately follow those discussing nucleic acid and lipid structure. The growth of molecular biology has led me to expand coverage of translation, gene regulation, recombinant DNA, transposons, the polymerase chain reaction, and chromatin and chromosomes. I have also added information on G proteins and signal transduction. I have made significant changes to the book in response to comments from readers and reviewers. I am particularly grateful to Neal Baer, James Baggott, and Mark O.J. Olson, who gave many helpful suggestions toward improving this edition.

P.J.F.

Biochemistry

Structure of Amino Acids

Structural biochemistry forms the foundation to the understanding of metabolic pathways. To dismiss the study of structure as unessential and dull is as senseless as refusing to learn anatomy. When we separate the chaff—in this case, the superfluous structural details—from the grain—the essential structural features—we are then left with a digestible foodstuff.

All living organisms contain α-amino acids, and all α-amino acids share the same structural backbone:

$$^+H_3N\!\!-\!\!-\!\!-\!\!-\!\!\overset{\overset{\textstyle R}{|}}{\underset{\underset{\textstyle H}{|}}{C}}\!\!-\!\!-\!\!-\!\!-\!\!C\!\!\overset{\textstyle O}{\underset{\textstyle O^-}{\diagdown}}$$

α-Amino group α-Carbon atom α-Carboxyl group

In other words, all α-amino acids possess three common features:

1. They have an α-carboxyl group. The α denotes that this group binds to the central or α-carbon atom, which is asymmetric.
2. They possess an α-amino group.
3. They contain a side chain, or R group, that is bound to the α-carbon.

In general, you need not memorize the exact structure of the R group for each of the twenty common, naturally occurring α-amino acids; knowing the proper classification is sufficient. You should also be familiar with the three-letter and single-letter abbreviations for each amino acid.

Each amino acid may be classified as **acidic, neutral,** or **basic,** depending on the charge on the R group at pH 7.0 (see Chap. 3). Acidic R groups bear a negative charge at pH 7.0 because they are strong proton donors. The two acidic α-amino acids—i.e., those with acidic R groups—are **aspartic acid** and **glutamic acid**:

Aspartic acid (Asp, D)

$$^-OOC\!\!-\!\!\overset{\beta}{C}H_2\!\!-\!\!\underset{\underset{\textstyle ^+NH_3}{|}}{\overset{\alpha}{C}H}\!\!-\!\!COO^-$$

β-Carboxyl

Glutamic acid (Glu, E)

$$^-OOC\!\!-\!\!\overset{\gamma}{C}H_2\!\!-\!\!\overset{\beta}{C}H_2\!\!-\!\!\underset{\underset{\textstyle ^+NH_3}{|}}{\overset{\alpha}{C}H}\!\!-\!\!COO^-$$

γ-Carboxyl

Glutamic acid differs from aspartic acid only in the number of CH_2 groups contained in its side chain. Each acid carries a charge of minus one at neutral pH.

The basic R groups of lysine and arginine carry a positive charge at pH 7.0. The imidazole group of histidine may be uncharged or positively charged at neutral pH. Because it can catalyze a variety of reactions, histidine is often found at the active site of enzymes.

Lysine (Lys, K)

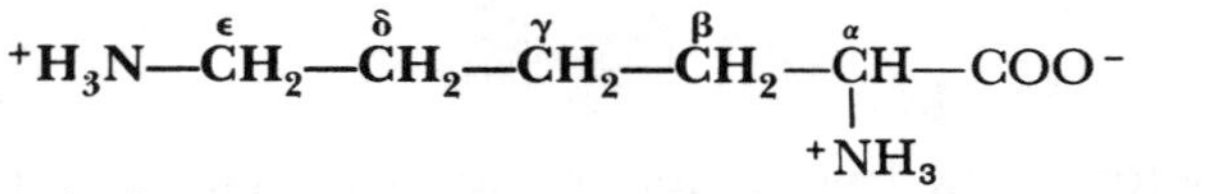

$$^+H_3N-\overset{\epsilon}{C}H_2-\overset{\delta}{C}H_2-\overset{\gamma}{C}H_2-\overset{\beta}{C}H_2-\overset{\alpha}{C}H-COO^-$$
$$\underset{^+NH_3}{|}$$

ε-Amino group

Arginine (Arg, R)

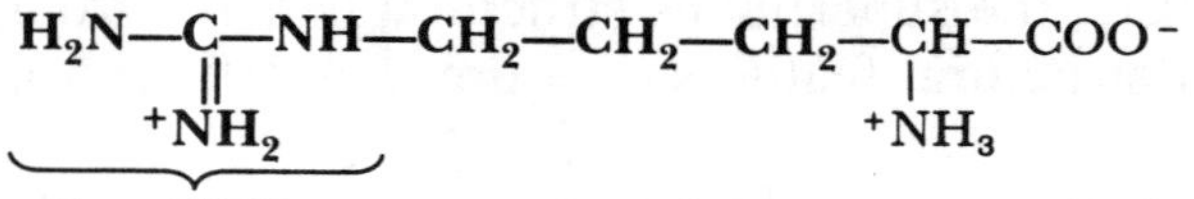

$$H_2N-\underset{\underset{+NH_2}{\|}}{C}-NH-CH_2-CH_2-CH_2-\underset{\underset{+NH_3}{|}}{CH}-COO^-$$

Guanidinium group

Histidine (His, H)

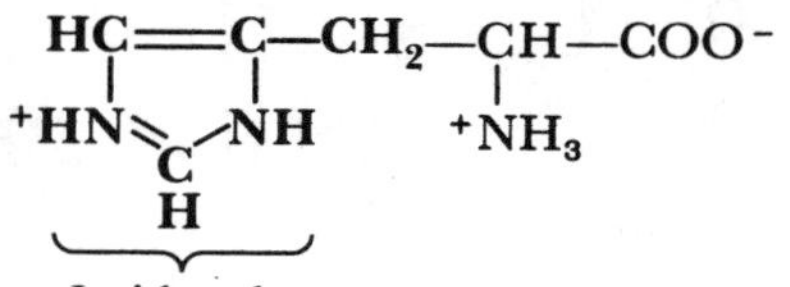

$$HC=C-CH_2-CH-COO^-$$

$$^+HN\diagdown\diagup NH \qquad ^+NH_3$$
$$\underset{H}{C}$$

Imidazole group

Two α-amino acids are commonly mislabeled as basic when in fact they are neutral: **glutamine** and **asparagine.** They are the amides of glutamic and aspartic acids, respectively. Although they are polar, the amide groups neither protonate nor dissociate.

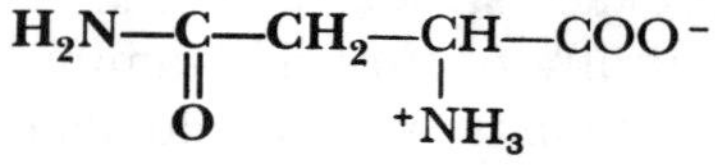

$$H_2N-\underset{\underset{O}{\|}}{C}-CH_2-\underset{\underset{+NH_3}{|}}{CH}-COO^- \qquad\qquad H_2N-\underset{\underset{O}{\|}}{C}-CH_2-CH_2-\underset{\underset{+NH_3}{|}}{CH}-COO^-$$

Asparagine (Asn, N) 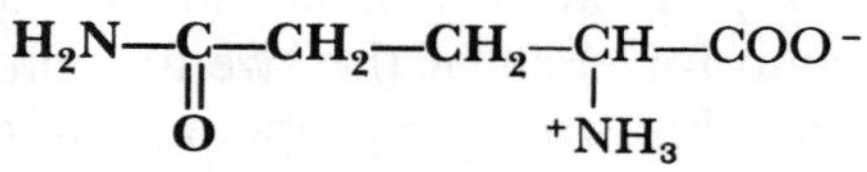   Glutamine (Gln, Q)

Amino acids are also classified according to whether or not they contain sulfur atoms, hydroxyl or aromatic groups, and branched or straight-chain hydrocarbons in their side chains (R groups). Each neutral amino acid may be further designated as **polar** or **nonpolar** according to the charge difference between different regions of its R group.

Cysteine and **methionine** contain sulfur. The —SH groups of cysteine can bind to one another to form the disulfide bridges that stabilize the structure of proteins. **Cystine** is a dimer of cysteine, in which two molecules of cysteine are joined via their sulfur atoms.

Methionine, as the name implies, has a methylated thiol group (sulfur atom), which is nonpolar.

Methionine (Met, M)

$$CH_3—S—CH_2—CH_2—\underset{\underset{^+NH_3}{|}}{CH}—COO^-$$

Cysteine (Cys, C)

$$HS—CH_2—\underset{\underset{^+NH_3}{|}}{CH}—COO^-$$

Sulfhydryl
group

Three α-amino acids contain aromatic groups: **phenylalanine, tyrosine,** and **tryptophan.** Phenylalanine consists of a phenyl ring bound to the methyl group of alanine. Tryptophan contains an indole group, which consists of a phenyl ring fused to a five-membered, nitrogen-containing ring. Both phenylalanine and tryptophan are nonpolar. The hydroxyl group of tyrosine, or *p*-hydroxyphenylalanine, renders it polar.

Phenylalanine (Phe, F)

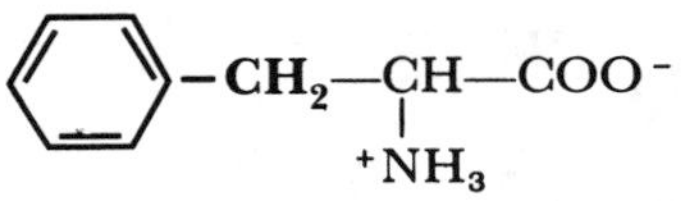

$$\text{⬡}—CH_2—\underset{\underset{^+NH_3}{|}}{CH}—COO^-$$

Tyrosine (Tyr, Y)

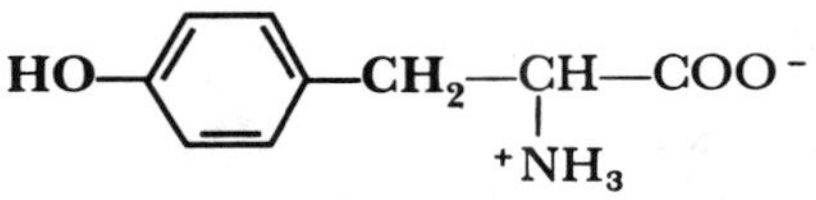

$$HO—\text{⬡}—CH_2—\underset{\underset{^+NH_3}{|}}{CH}—COO^-$$

Tryptophan (Trp, W)

$$\text{⬡}\underset{\underset{H}{\overset{|}{N}}}{\overset{\overset{C—CH_2—\underset{\underset{^+NH_3}{|}}{CH}—COO^-}{\|}}{{}}}CH$$

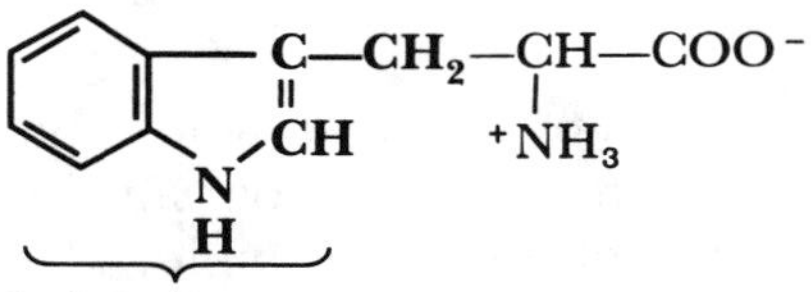

Indole group

Three α-amino acids have branched hydrocarbon chains: **leucine, isoleucine,** and **valine.** Since their side chains are purely hydrocarbons, they are therefore nonpolar.

$$CH_3—\underset{\underset{CH_3}{|}}{CH}—CH_2—\underset{\underset{^+NH_3}{|}}{CH}—COO^-$$

Leucine (Leu, L)

$$CH_3—CH_2—\underset{\underset{CH_3}{|}}{CH}—\underset{\underset{^+NH_3}{|}}{CH}—COO^-$$

Isoleucine (Ile, I)

$$CH_3—\underset{\underset{CH_3}{|}}{CH}—\underset{\underset{^+NH_3}{|}}{CH}—COO^-$$

Valine (Val, V)

The R groups of **alanine** and **glycine** are a methyl group and a hydrogen atom, respectively. The methyl side chain of alanine makes it nonpolar.

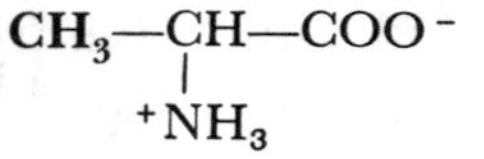

$$CH_3—CH—COO^-$$
$$|$$
$$^+NH_3$$

Alanine (Ala, A)

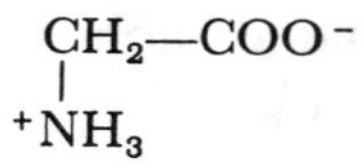

$$CH_2—COO^-$$
$$|$$
$$^+NH_3$$

Glycine (Gly, G)

The R groups of **serine** and **threonine** contain hydroxyl groups, like tyrosine, which render them polar.

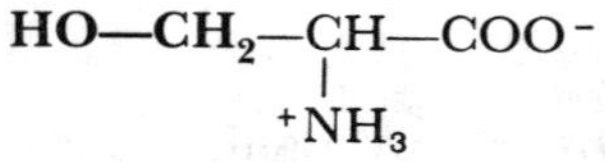

$$HO—CH_2—CH—COO^-$$
$$|$$
$$^+NH_3$$

Serine (Ser, S)

$$CH_3—CH—CH—COO^-$$
$$| |$$
$$OH ^+NH_3$$

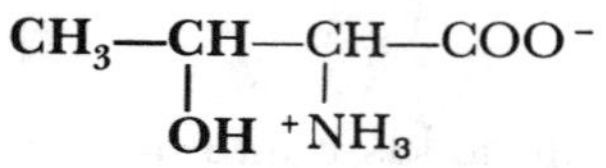

Threonine (Thr, T)

Proline is an imino rather than an amino acid because it contains a secondary rather than a primary amino group. Because of the rigidity of its five-membered ring, proline residues will kink a chain of amino acids. Proline is nonpolar.

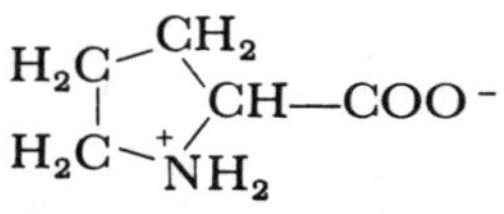

$$H_2C{-}^{CH_2}$$
$$| \quad CH—COO^-$$
$$H_2C{-}^+NH_2$$

Proline (Pro, P)

Ninhydrin reacts with the free α-amino groups of amino acids and proteins to produce a purple color. The ninhydrin reaction can be used to estimate the quantity of amino acid present in a sample. The quantitation of individual amino acids involves their separation by chromatographic techniques.

Problems
Problems 1–3
Match the structures below to the descriptions in Problems 1–3. Each structure may be used more than once or not at all.

A. 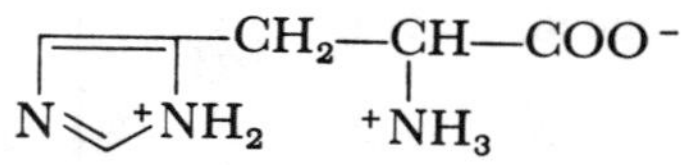
$$—CH_2—CH—COO^-$$
$$N{\diagdown}^+NH_2 \quad ^+NH_3$$

B. 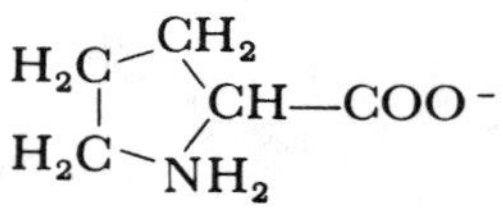
$$H_2C{-}^{CH_2}$$
$$| \quad CH—COO^-$$
$$H_2C{-}NH_2$$

C. 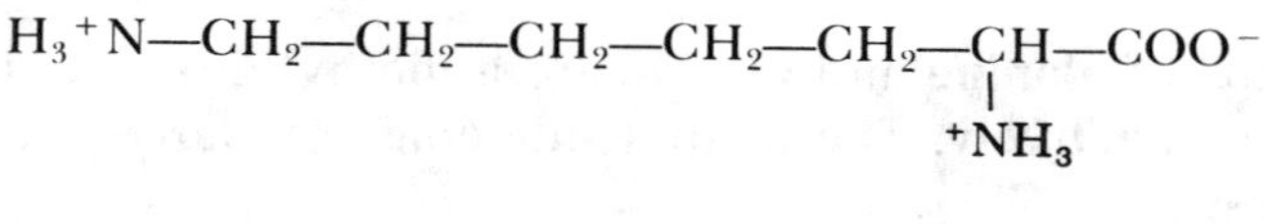 $$H_3{}^+N—CH_2—CH_2—CH_2—CH_2—CH_2—CH—COO^-$$
$$|$$
$$^+NH_3$$

D. $^-OOC—CH_2—\underset{\underset{+NH_3}{|}}{CH}—COO^-$

E. $H_2N—\underset{\underset{O}{\|}}{C}—CH_2—\underset{\underset{+NH_3}{|}}{CH}—COO^-$

1. Acidic amino acid.   D
2. Often responsible for kinks in polypeptide chain.   B
3. Despite the amino group in its side chain, this amino acid is neutral rather than basic.   E

Answers

1. D. This is aspartic acid.
2. B. This is proline.
3. E. This is asparagine.

References

Devlin, T. M. *Textbook of Biochemistry with Clinical Correlations* (3rd ed.). New York: Wiley-Liss, 1992. Pp. 25–31.

Mathews, C. K., and van Holde, K. E. *Biochemistry*. Redwood City, Calif.: Benjamin/Cummings, 1990. Pp. 133–141.

Murray, R. K., Granner, D. K., Mayes, P. A., and Rodwell, V. W. *Harper's Biochemistry* (22nd ed.). Norwalk, Conn.: Appleton & Lange, 1990. Pp. 21–31.

Stryer, L. *Biochemistry* (3rd ed.). New York: Freeman, 1988. Pp. 15–21.

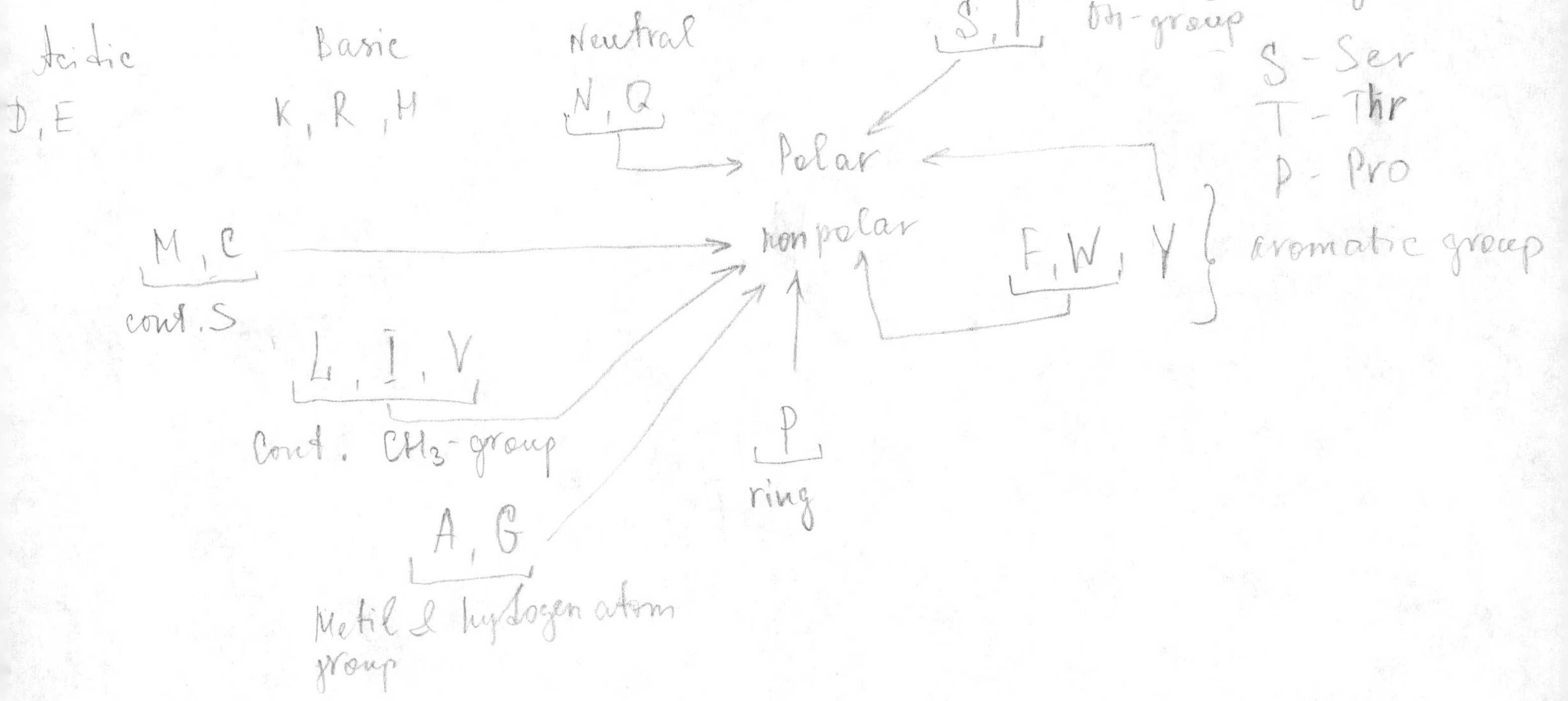

Handwritten annotations:

D – Asp
E – Glu
K – Lys
R – Arg
H – His
N – Asn
Q – Gln
M – Met
C – Cys
F – Phe
W – Trp
Y – Tyr
L – Leu
I – Ile
V – Val
A – Ala
G – Gly
S – Ser
T – Thr
P – Pro

acidic: D, E
Basic: K, R, H
Neutral: N, Q
S, T OH-group
M, C cont. S
L, I, V cont. CH₃-group
A, G Metil & hydrogen atom group
P ring
F, W, Y aromatic group
Polar / nonpolar

Structure and Properties of Polypeptides and Proteins

Peptide Bonds

Peptide bonds, a type of amide bond, weld the amino end (N-terminal) of one amino acid to the carboxyl end (C-terminal) of another, as shown below. Dehydration forges this peptide linkage.

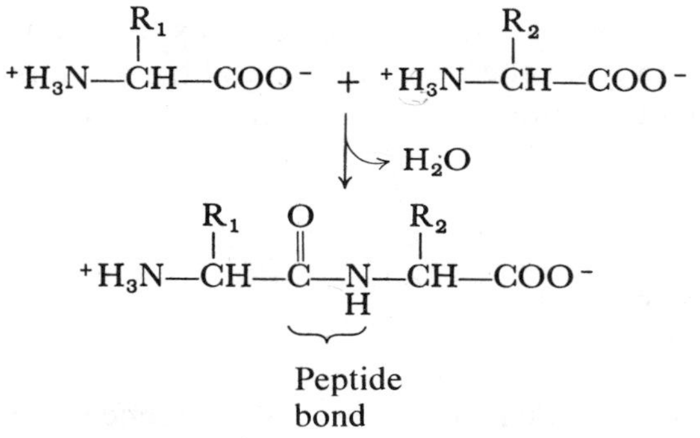

$$^+H_3N—\underset{\underset{R_1}{|}}{CH}—COO^- \; + \; ^+H_3N—\underset{\underset{R_2}{|}}{CH}—COO^-$$

$$\downarrow \; H_2O$$

$$^+H_3N—\underset{\underset{R_1}{|}}{CH}—\underset{\overset{O}{\|}}{C}—\underset{\underset{H}{|}}{N}—\underset{\underset{R_2}{|}}{CH}—COO^-$$

Peptide
bond

The C—N bond in the peptide linkage has partial double-bond properties that make it rigid and prevent the adjacent groups from rotating freely.

Neither the C=O nor the N—H in the peptide bond can dissociate.

By convention, the N-terminus is shown to the left of the C-terminus.

Each amino acid in a polypeptide chain is termed a **residue.**

Proteolysis

Peptide-bond hydrolysis, or **proteolysis,** requires either the presence of proteolytic enzymes or heating at 110°C for 24 hours in the presence of 6 N HCl (acid hydrolysis) or NaOH (alkaline hydrolysis). Acid hydrolysis, unfortunately, destroys tryptophan and cysteine and partially destroys serine, threonine, and tyrosine. Alkaline hydrolysis destroys cysteine, serine, and threonine. Since it does not damage tryptophan, alkaline hydrolysis can be used in quantitative determinations of this amino acid.

7

Amino Acid Composition of Polypeptides

The distinction between the terms **oligopeptide** and **polypeptide** is somewhat arbitrary. An amino acid chain with less than 25 amino acids is called an oligopeptide, whereas a polypeptide has more than 25 amino acids. A **protein** may consist of a long polypeptide or several polypeptide subunits.

The first step in determining the amino acid composition of a polypeptide is to measure the molecular weight of the whole, purified compound. Next, the percentage composition of each amino acid is determined, usually by automated amino acid analysis. Automated amino acid analyzers contain ion-exchange chromatography columns that separate free amino acids after complete acid hydrolysis of the polypeptide. Each amino acid is eluted from the column at a characteristic pH and is quantitated spectrophotometrically by measuring the optical density after adding ninhydrin. A second sample, which is subjected to alkaline hydrolysis, is run through the amino acid analyzer to determine the tryptophan context. Alternatively, direct spectrophotometric tests could be used for tryptophan, which has an absorption peak in the ultraviolet range.

Finally, from the percentage composition, one calculates the number of units of each amino acid from the molecular weight, and thus the exact quantitative composition may be determined.

Structure of Polypeptides
Primary Structure

The primary structure of a polypeptide is its amino acid sequence and the location of disulfide bonds, if present. The principal method for determining primary structure is the Edman reaction. In step 1 of the Edman reaction, phenyl isothiocyanate combines with the terminal amino acid of the polypeptide. In step 2, the N-terminal peptide bond is then cleaved, yielding a phenylthiohydantoin (PTH) amino acid, which can be identified by chromatography. In the process, the polypeptide chain is shortened by one residue. The cycle of the Edman reaction is then repeated, each time identifying the new N-terminal amino acid of the polypeptide.

The Edman reaction is not reliable for polypeptide chains with more than 50 residues. To analyze the primary structure of such longer chains, one must first cleave them into shorter fragments. This cleavage can be achieved with either chemicals or enzymes. Cyanogen bromide (CNBr) cleaves peptide chains at the carboxyl end of methionine. **Trypsin** cleaves the peptide bond at the carboxyl end of the two strongly basic amino acids: arginine and lysine. **Chymotrypsin** cleaves the peptide bond at the carboxyl end of the three aromatic amino acids: phenylalanine, tyrosine, and tryptophan. Pancreatic enzymes like trypsin and chymotrypsin are stored and secreted as inactive proenzymes. On entering the duodenum, they become activated.

By using a variety of reagents separately on a polypeptide, different sets of oligopeptides are generated. The individual amino acid sequences of these fragments overlap and can be fitted together like pieces of a jigsaw puzzle.

To sequence proteins composed of two or more polypeptide chains, the noncovalent bonds between chains must be cleaved by denaturing agents such as urea. The chains are then separated and sequenced individually.

To sequence polypeptide chains containing disulfide bonds, performic acid is used to break the disulfide bonds. Changes in electrophoretic mobility after cleavage identify the cysteine residues involved in the disulfide bonds.

An alternative method to determine the primary structure of a polypeptide is to sequence the DNA coding for the polypeptide. This method elucidates the primary structure of the newly synthesized protein but does not account for post-transcriptional or post-translational modifications (described in later chapters).

Once the primary structure of a polypeptide is known, one can compare its structure to those of other polypeptides in an attempt to understand its function. Internal repetition of amino acid sequences is common in a number of polypeptides, particularly the immunoglobulins. DNA probes can be synthesized to search for the gene producing the polypeptide.

Secondary Structure

The spatial arrangements of amino acid residues close to one another in the linear sequence of a polypeptide chain are termed **secondary structures**. The most common secondary structures are the α-helix, the β-pleated sheet, the random coil, and the triple helix.

The α-helix forms spontaneously. Each turn in the spiral of the helix contains 3.6 amino acid residues. The helix is stabilized by hydrogen bonding between the CO and NH group situated four residues ahead in the same polypeptide chain. Two or more α-helices can combine, creating rope-like proteins such as fibrin and keratin.

The polypeptide chain in β-pleated sheets is almost fully extended, in contrast to the tight coil of the α-helix. Two or more polypeptide chains can fuse into a larger pleated sheet, which is stabilized by hydrogen bonding between the CO and NH groups on different polypeptide chains. Adjacent polypeptide chains may run in the same direction (parallel structure) or in the opposite direction (antiparallel structure).

Sections of polypeptides with few regularities in structure are termed **random coils**.

Collagen is a unique fibrous protein built from tropocollagen. Tropocollagen has an unusual amino acid composition. Nearly one-third of its residues are glycine, which are often spaced every third residue in the chain. The proline content of tropocollagen is very high and two unusual amino acids are present: hydroxyproline and hydroxylysine.

Tropocollagen consists of three polypeptide chains, each helical, entwined into a long cable. Collagen is built from an array of tropocollagen molecules placed end to end and parallel to one another. The staggered positioning of the ends of the tropocollagen molecules creates the striations in collagen seen under electron microscopy.

Unlike collagen, most proteins are globular in shape. To form such compact proteins, polypeptide chains must change directions to allow infolding. The β-turn is a common way to change the directions of polypeptide chains. Glycine and proline are often found at such tight bends.

Tertiary Structure

The spatial arrangement of amino acid residues widely separated in the linear sequence of a polypeptide chain is termed its **tertiary structure**. Thus, tertiary structure refers to the overall three-dimensional structure of a polypeptide chain. In general, proteins may have either of two tertiary forms: fibrous or globular. The primary structure of a polypeptide determines its secondary and tertiary structure.

The **fibrous proteins** are elongated and water-insoluble. Their secondary structure is generally either the α-helical, pleated-sheet, or triple-helical forms, rather than random coils. Fibrous proteins include collagen, elastin, and fibronectin.

In contrast, the **globular proteins** are ellipsoidal and water-soluble, and they consist mainly of random coils with occasional stretches of α-helices. The tertiary structure of globular proteins, however, is far from random, inasmuch as the primary structure of any polypeptide determines its tertiary structure. The nonpolar or hydrophobic amino acids—such as alanine, leucine, and tryptophan—tend to fold into the central area of a globular protein to exclude water as much as possible. Furthermore, these nonpolar side chains weakly attract one another. Polar and ionized side chains, on the other hand, tend to move toward the outer protein surface to form hydrogen bonds with water; they are electrostatically attracted toward side chains of opposite charge and repelled from those of like charge.

The partial double-bond nature of the peptide linkages restricts the folding of the polypeptide, and the size of the R groups also helps to govern tertiary structure. Small side chains allow tight folding of the chain, whereas bulky R groups prevent the close approach of other groups. Disulfide bonds stabilize the tertiary structure of proteins by rigidly linking cysteine residues to one another, forming cystine.

Denaturation, or unfolding, of globular proteins occurs after heating or treatment with strong acids, strong bases, concentrated urea, or other agents. Proteins lose their activity after denaturation, although under certain conditions, this loss may be reversible.

Quaternary Structure

The arrangement of polypeptide chains in relation to one another in a multiple-chained protein is called the **quaternary structure.** The bonds linking these chains are all noncovalent, such as hydrogen bonds, electrostatic (salt) bonds, and hydrophobic bonds.

Oxygen-Carrying Proteins

Oxygen is poorly soluble in water. Two oxygen-carrying proteins, hemoglobin in red blood cells and myoglobin in muscle, play a vital role in oxygen transport. The oxygen-binding unit of hemoglobin and myoglobin is heme. Heme is a prosthetic group. Prosthetic groups of proteins are the nonpolypeptide components. Heme consists of a protoporphyrin bound to iron.

Hemoglobin is built from four polypeptide chains linked noncovalently. Each chain has its own heme group with a single oxygen-binding site. The principal adult hemoglobin, hemoglobin A, has two α- and two β-chains. Hemoglobin A_2, a minor hemoglobin in adults, has two α- and two δ-chains.

Unlike hemoglobin, myoglobin consists of a single polypeptide chain. Although myoglobin has a markedly different primary structure than hemoglobin, the tertiary structure of its main chain closely resembles that of the α- and β-chains of hemoglobin. The binding of one oxygen molecule to myoglobin does not influence subsequent oxygen binding to other myoglobin molecules. In contrast, the binding of the first oxygen molecule to deoxyhemoglobin facilitates subsequent oxygen binding to other heme groups of the same hemoglobin molecule. This facilitation is known as cooperative binding and causes the sigmoidal shape in the oxygen dissociation curve for hemoglobin (Fig. 2-1). Hemoglobin binds oxygen cooperatively because it is an allosteric protein: a protein that changes conformation in response to certain substances termed allosteric effectors. Allosteric proteins all have multiple subunits. The oxygen affinity of hemoglobin but not myoglobin depends on pH, P_{CO_2}, and 2,3-bisphosphoglycerate (BPG). Lowering pH shifts the oxygen dissociation curve of hemoglobin to the right, as shown in Figure 2-1.

In addition to carrying oxygen, hemoglobin also carries CO_2 as carbamates. CO_2 reacts with the amino groups of hemoglobin to form carbamates:

$$CO_2 + R-NH_2 \rightleftharpoons R-\overset{\overset{\displaystyle H}{|}}{N}-\overset{\overset{\displaystyle O}{\|}}{C}-O^- + H^+$$

Carbamates, in turn, form salt bridges to stabilize deoxyhemoglobin. A rise in P_{CO_2} shifts the curve to the right, leading to reduced oxygen binding. Thus the local accumulation of H^+ and CO_2 in metabolically active tissues leads to improved oxygen delivery from hemoglobin.

One molecule of BPG binds to each hemoglobin. BPG cross-links the β-chains, stabilizing the quaternary structure of deoxyhemoglobin. BPG markedly lowers the oxygen affinity of hemoglobin, thereby promoting oxygen delivery to tissues. Oxygen binding to hemoglobin

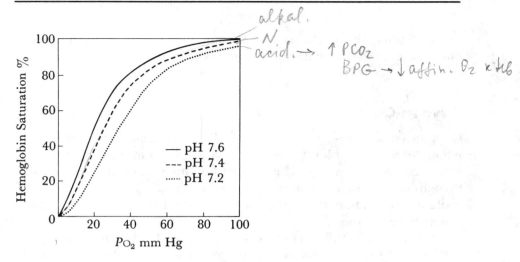

Fig. 2-1 Oxygen dissociation curves of hemoglobin.

depends on breaking salt links between the four polypeptide chains. More links must be broken for binding of the first oxygen molecule than for the second, third, or fourth. As a result of this cooperative oxygen binding, hemoglobin is a much better oxygen carrier than myoglobin.

Problems

Problem 1

Total acid hydrolysis of a pentapeptide complemented by total alkaline hydrolysis yields an equimolar mixture of five amino acids (listed alphabetically): Ala, Cys, Lys, Phe, Ser. N-terminal analysis with the Edman reaction generates PTH-serine. Trypsin digestion produces a tripeptide whose N-terminal residue is Cys and a dipeptide with Ser at its N-terminal. Chymotrypsin digestion of the above tripeptide yields Ala plus another dipeptide. Choose the *incorrect* statement.

    A. The dipeptide from trypsin digestion is Ser-Lys.
    B. The tripeptide from trypsin digestion is Cys-Phe-Ala.
    C. The pentapeptide is Ser-Lys-Cys-Phe-Ala.
    D. Cyanogen bromide would be useful for sequencing the pentapeptide.

Problem 2

In a globular protein, which amino acid is most likely to be positioned deep within the molecule?

    A. Aspartic acid.
    B. Lysine.
    C. Glutamic acid.
    D. Leucine.
    E. Arginine.

Problem 3

The oxygen affinity of hemoglobin is not increased by:

A. A decrease in BPG.
B. Alkalosis.
C. Low P_{CO_2}.
D. The binding of the first oxygen molecule to deoxyhemoglobin.
E. Cyclic AMP.

Problem 4

Choose the single best statement about collagen:

A. Collagen has an unusually high leucine content.
B. The basic subunit consists of three tropocollagen fibers joined in a triple helix.
C. Collagen does not have an α-helical secondary structure.
D. The proline content of collagen is quite low.
E. Collagen has large segments of β-pleated sheets.

Problem 5

The oxygen affinity of myoglobin:

A. Exceeds that of hemoglobin.
B. Is increased by a drop in pH.
C. Does not change when P_{CO_2} drops.
D. Is reduced by a rise in 2,3-bisphosphoglycerate.
E. Improves in an allosteric manner following oxygen binding.

Problem 6

Sickle cell hemoglobin (hemoglobin S) differs in primary structure from hemoglobin A in only one residue: valine instead of glutamate at position 6 of the β-chain. Choose the single best statement about hemoglobin S:

A. Hemoglobin S has the same electrophoretic mobility as hemoglobin A.
B. In theory hemoglobin S should have the same secondary and tertiary structures as hemoglobin A because valine has similar properties to glutamate.
C. In homozygous form, hemoglobin S leads to a more fibrous deoxyhemoglobin than occurs with hemoglobin A.
D. Hemoglobin S has a grossly abnormal structure in both its oxy- and deoxy- forms.

Answers

1. D. Cyanogen bromide cleaves at the carboxyl end of methionine. This pentapeptide lacks methionine. Hence cyanogen bromide would not assist in sequencing this peptide.

2. D. As a nonpolar or hydrophobic amino acid, leucine is likely to be situated within the protein. The other choices are polar.
3. E.
4. C. Glycine, hydroxylysine, and hydroxyproline are present in high proportions. Tropocollagen, not collagen, is a triple helix. The helical structure of the polypeptide chains of tropocollagen differs from an α-helix because the spiral is tighter.
5. C. Unlike hemoglobin the oxygen affinity of myoglobin is not influenced by pH, P_{CO_2}, BPG, or the binding of the first oxygen molecule to the deoxy form of the protein. Myoglobin consists of a single polypeptide chain and cannot therefore act as an allosteric protein as can hemoglobin.
6. C. Glutamate has a negatively charged side chain, whereas that of valine is nonpolar. Hence the two proteins will differ in charge and therefore in mobility on electrophoresis. Substituting a nonpolar amino acid for a polar one could be expected to impact on secondary and tertiary structure. The glutamate residue in question on the β-chain of hemoglobin A is on the outside of the molecule. Substituting valine creates an abnormally adhesive site on the outside of hemoglobin S. When deoxyhemoglobin S predominates over oxyhemoglobin S, the molecules stick together, forming long fibrous strands. When oxyhemoglobin predominates, there is little or no aggregation of hemoglobin S.

References

Devlin, T. M. *Textbook of Biochemistry with Clinical Correlations* (3rd ed.). New York: Wiley-Liss, 1992. Pp. 31–34, 38–67, 72–88.

Mathews, C. K., and van Holde, K. E. *Biochemistry*. Redwood City, Calif.: Benjamin/Cummings, 1990. Pp. 142–150, 162–205, 216–248.

Murray, R. K., Granner, D. K., Mayes, P. A., and Rodwell, V. W. *Harper's Biochemistry* (22nd ed.). Norwalk, Conn.: Appleton & Lange, 1990. Pp. 32–57.

Stryer, L. *Biochemistry* (3rd ed.). New York: Freeman, 1988. Pp. 22–70, 143–173, 261–281.

Acids, Bases, and Buffers

In biochemistry the most workable definitions of acids and bases are those of Brønsted, who defined an **acid** as a proton donor and a **base** as a proton acceptor. For each acid and each base, there is its **conjugate base** and **conjugate acid**, respectively, from which it differs by the proton lost or gained.

The Henderson-Hasselbalch Equation

Let RH represent an acid and R⁻ its conjugate base. Its dissociation may be represented by:

$$RH \rightleftharpoons H^+ + R^-$$

The ionization, or dissociation, constant of this acid, K_a, is defined by the equilibrium expression:

$$K_a = \frac{[H^+][R^-]}{[RH]} \tag{3-1}$$

where the square brackets indicate the molar concentrations of the substances. By rearrangement and substitution utilizing the definitions $pH = -\log[H^+]$ and $pK_a = -\log K_a$, we get:

$$[H^+] = \frac{K_a[RH]}{[R^-]}$$

$$-\log[H^+] = -\log K_a + \log \frac{[R^-]}{[RH]}$$

$$pH = pK_a + \log \frac{[R^-]}{[RH]} \tag{3-2}$$

Note that $\log A \cdot \left(\frac{B}{C}\right) = \log A + \log \left(\frac{B}{C}\right) = \log A - \log \left(\frac{C}{B}\right)$

The pK_a is the pH at which half of the acid is dissociated. Equation 3-2 is the **Henderson-Hasselbalch equation,** which may also be written:

$$pH = pK_a + \log \frac{[\text{proton acceptor}]}{[\text{proton donor}]}$$

If the pH differs by 2 or more units from the pK_a, you will not, for most practical purposes, need to use the Henderson-Hasselbalch equation to calculate the concentrations of the components of the dissociation reaction, because 99% or more of the substance will exist as the proton acceptor (if pH $\geq pK_a + 2$) or the proton donor (if pH $\leq pK_a - 2$).

The meaning of the Henderson-Hasselbalch equation can be easily conceptualized. If the pH drops below the pK_a, the conjugate base (R^-) is protonated to the acid (RH); hence, the ratio $[R^-]/[RH]$ falls below 1. On the other hand, if the pH rises above the pK_a, the acid (RH) liberates its proton and the $[R^-]/[RH]$ ratio rises above 1.

Titration and Buffers

Titration is the incremental addition of a strong acid or base to a solution while measuring its pH up to the point, say, of neutralization. After the desired pH is reached, one calculates the moles of acid or base added, and from that figure, one determines the quantity of titratable acid or base in the solution.

The results of titration demonstrate whether or not the substance in solution is acting as a **buffer,** that is, a combination of a weak acid and its salt that changes pH relatively slowly in response to the addition of strong acid or base. Most buffers exhibit their buffering action within a narrow pH range ($pK_a \pm 1$).

An important application of titration to medicine is in renal physiology. The **titratable acidity** of urine is defined as the number of millimoles of NaOH required to titrate 1 liter of urine up to physiologic pH (7.4). The principal titratable acid found in the urine is phosphate, which exists in four different forms and has two pK_a values:

$$H_3PO_4 \xrightleftharpoons[\;]{pK_{a1} = 2.1} H^+ + H_2PO_4^- \xrightleftharpoons[\;]{pK_{a2} = 6.8} H^+ + HPO_4^{-2} \rightleftharpoons PO_4^{-3}$$

Since the pH of urine never drops below 4.5, there is virtually no H_3PO_4 in urine, because $pK_{a1} = 2.1$ (recall the value of $[R^-]/[RH]$ when pH $\geq pK_a + 2$).

Amino Acids as Buffers

Proteins function as one of the most important buffer systems in blood and tissues, and their buffering ability derives from the dissociable groups on their constituent amino acids.

Glycine, the simplest amino acid, has two dissociable groups, the α-amino and α-carboxyl groups:

$^+H_3N—CH_2—COOH$

$\updownarrow$ $pK_{a_1} = 2.3$

$^+H_3N—CH_2—COO^-$

$\updownarrow$ $pK_{a_2} = 9.6$

$H_2N—CH_2—COO^-$

Glycine can therefore exist in three forms, depending on the pH:

1. Completely protonated ($^+H_3N—$, $—COOH$). Net charge of $+1$.
2. Protonated α-amino group with unprotonated α-carboxyl group ($^+H_3N—$, $—COO^-$). This is the **isoelectric** species of glycine because it has 0 net charge.
3. Completely unprotonated ($H_2N—$, $—COO^-$). Net charge of -1.

To determine the form and net charge of glycine at pH 3.0, for example, consider each dissociable group separately. Since pK_{a_1} for the α-carboxyl group is 2.3, this group will exist mainly as COO^-. More than about 10%, however, will be in the COOH form, because $pH - pK_a < 1.0$. Applying the Henderson-Hasselbalch equation,

$3.0 = 2.3 + \log[COO^-]/[COOH]$

$[COO^-]/[COOH] = 5$

Therefore, $\frac{5}{6}$ of the carboxyl groups will exist as COO^-, while $\frac{1}{6}$ will be COOH. The net charge due to the carboxyl groups will be $-\frac{5}{6}$.

Turning to the α-amino group, its pK_{a_2} of 9.6 is 6.6 units above pH 3.0. You do not need to use the Henderson-Hasselbalch equation here, because virtually 100% of this group will exist as $^+NH_3$ with a charge of $+1$. Thus, the net charge on glycine at pH 3.0 is $+\frac{1}{6}$.

The net charge on an amino acid or protein determines its mobility in **electrophoresis,** a technique that is used to separate compounds of differing charge by applying an electrical potential across a medium, such as paper or starch gel, that contains a solution of buffer and the sample compounds. Since glycine at pH 3.0 has a charge of $+\frac{1}{6}$, it will move toward the negative pole during electrophoresis at this pH.

Because its imidazole group has a pK_a of 6.0, histidine is the only amino acid with significant buffering capacity in the physiologic range, i.e., between pH 6.0 and pH 8.0. The pK_a of cysteine's sulfhydryl group is 8.3, which is closer to the normal blood pH of 7.44 than the pK_a of the imidazole group, but this SH group is often tied up in disulfide linkages, such as in the dimer cystine, and is therefore unable to act as a buffer.

Let us next consider the buffering properties of histidine:

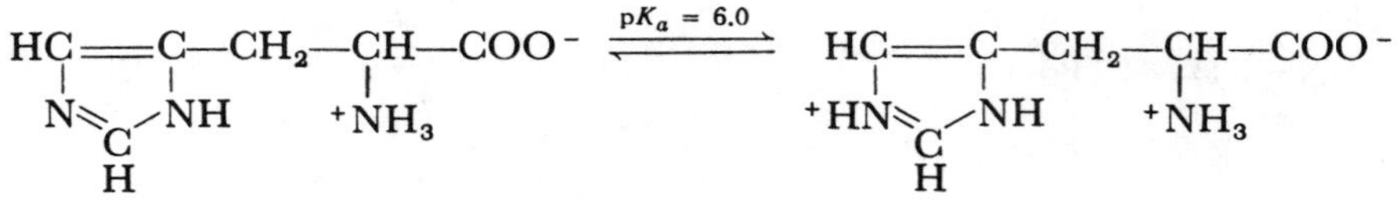

$$HC{=\!\!=}C{-}CH_2{-}CH{-}COO^- \xrightarrow{pK_a = 6.0} HC{=\!\!=}C{-}CH_2{-}CH{-}COO^-$$

(imidazole structures with N, NH, C—H ring and $^+NH_3$ side chain)

At pH 7.4 the α-carboxyl group ($pK_a = 1.8$) will exist as COO^-, while the α-amino group ($pK_a = 9.0$) will exist almost entirely as $^+NH_3$. For the imidazole group,

$$7.4 = 6.0 + \log \frac{[N\!\!\diagup]}{[^+HN\!\!\diagup]}$$

$$\frac{[N\!\!\diagup]}{[^+HN\!\!\diagup]} = 25$$

Hence, at this pH, the charge due to this group is $+\frac{1}{26}$. Since the -1 charge on the α-carboxyl group balances the $+1$ charge on the α-amino group, the net charge on histidine at pH 7.4 is $+\frac{1}{26}$.

The Bicarbonate Buffer

Bicarbonate (HCO_3^-) provides the most important physiologic buffer system for several reasons:

1. Bicarbonate is present in high concentration in plasma; e.g., in humans, its concentration is 25 mEq/liter at pH 7.4 when $P_{CO_2} = 40$ mm Hg.
2. The enzyme carbonic anhydrase in erythrocytes (red blood cells) and renal tubules allows the rapid interconversion of CO_2 and carbonic acid (H_2CO_3).
3. The lungs regulate the partial pressure of CO_2 (P_{CO_2}) in the blood on a minute-by-minute basis.
4. The kidneys regulate urinary bicarbonate excretion and can alter the plasma HCO_3^- concentration slowly, over the course of hours.
5. Hemoglobin is positioned side-by-side with carbonic anhydrase in erythrocytes and assists the buffering action of bicarbonate.

Let us first consider the bicarbonate buffer system in vitro, unassisted by the lungs, kidneys, carbonic anhydrase, or hemoglobin.

$$CO_2 + H_2O \rightleftharpoons H_2CO_3 \xrightarrow{pK_a = 6.1} H^+ + HCO_3^-$$

In vitro, the hydration of CO_2 to form carbonic acid, H_2CO_3, and the reverse reaction occur very slowly. As a buffer at pH 7.4, this system compares in strength to the histidine buffer. If we add HCl to this buffer, the rapid net reaction will be:

$$H^+ + HCO_3^- \longrightarrow H_2CO_3 \longrightarrow CO_2 + H_2O$$

When HCl is added to bicarbonate in the presence of the enzyme carbonic anhydrase, the rapid net reaction will be:

$$H^+ + HCO_3^- \longrightarrow H_2CO_3 \xrightarrow{\text{Carbonic anhydrase}} CO_2 + H_2O$$

Applying the Henderson-Hasselbalch equation, we get:

$$pH = 6.1 + \log \frac{[HCO_3^-]}{[H_2CO_3]}$$

In the presence of carbonic anhydrase, the H_2CO_3 concentration will be so small that it cannot be accurately measured. Instead, we may substitute the CO_2 concentration (mmoles/liter):

$$pH = 6.1 + \log \frac{[HCO_3^-]}{[CO_2]}$$

In practice, we measure the partial pressure of carbon dioxide (P_{CO_2} in mm Hg or torr), rather than measuring the CO_2 concentration directly:

$$[CO_2] \text{ (mmoles/liter)} = 0.03 P_{CO_2} \text{ (mm Hg)}$$

Therefore,

$$pH = 6.1 + \log \frac{[HCO_3^-]}{0.03 P_{CO_2}}$$

Adding HCl to this system consumes HCO_3^-, thereby lowering the HCO_3^- concentration and raising the P_{CO_2}. Because it is volatile, some of this CO_2 will enter the air and lower the amount of dissolved CO_2.

Now let us consider the in vivo bicarbonate buffer system that utilizes the lungs and kidneys, but, hypothetically, lacks hemoglobin and other buffers. If we add HCl, we will rapidly lower the HCO_3^- concentration and raise the P_{CO_2}. Immediately, the lungs will compensate by increasing the ventilation through augmenting the tidal volume, the respiratory rate, or both, to boost the CO_2 loss from the lungs and lower blood Pco_2. Over a period of several hours, the kidneys will increase their tubular reabsorption of bicarbonate to restore slowly the serum HCO_3^- level. Thus, the renal control of urinary bicarbonate excretion along with the pulmonary control of CO_2 exchange make this in vivo buffer system much more powerful in the long run than the other physiologic buffer systems.

Problems

Problems 1–2
Match the choices below to Problems 1–2.

    A. Bicarbonate buffer.
    B. Histidine buffer.
    C. Phosphate buffer.
    D. Carboxyl group of amino acids.

1. Does not buffer within physiologic pH range.
2. Carbonic anhydrase inhibitors make this buffer less effective.

Problem 3
Using the pK_a values listed below, identify amino acids *A*, *B*, *C*, and *D* from their titration curves in Figure 3-1.

Amino acid	pK_{a_1}	pK_{a_2}	pK_{a_3}
Asp	2.0	3.9	10.0
His	1.8	6.0	9.2
Lys	2.2	9.2	10.8
Thr	2.6	10.4	

Fig. 3-1 Titration curves for amino acids in Problem 3.

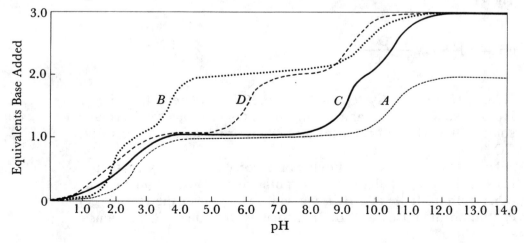

Problems 4–5
Match the [R$^-$]/[RH] ratios below to the pH values listed in Problems 4–5, using the Henderson-Hasselbalch equation.

   A. 1.0
   B. 10.0
   C. 0.10
   D. 0.01

4. pH = pK_a − 2
5. pH = pK_a + 1

Problems 6–7
A sample of aspartic acid is titrated from pH 1.0 to pH 6.5 by the addition of 3.0 mmoles of KOH. The pK_a values for aspartic acid are 2.0 (α-carboxyl group), 3.9 (β-carboxyl group), and 10.0 (α-amino group).

6. How many mmoles of aspartic acid are present in the sample?
7. Choose the single best answer:
   A. Both carboxyl groups and the amino group will act as buffers during this titration.
   B. Aspartic acid is not an effective buffer within this pH range.
   C. The amide group of aspartic acid does not contribute to its ability to act as a buffer.
   D. None of the above.

Answers

1. Carboxyl groups of amino acids are effective buffers only at low pH (1.0–4.0).
2. A.
3. Curve A is threonine, B is aspartic acid, C is lysine, and D is histidine.
4. D. Let x = [R$^-$]/[RH]. pH = pK_a + log x. When pH = pK_a −2, log x = −2 and x = 0.01.
5. B. When pH = pK_a + 1, log x = 1 and x = 10.
6. Starting at pH 1.0 the KOH must titrate about 90% of the α-carboxyl group to reach pH 6.5. It must also titrate nearly 100% of the β-carboxyl group. Hence the mmoles of aspartic acid equals 3.0/1.9 or 1.6.
7. D. The amino group will not act as a buffer during this titration. Aspartic acid is an effective buffer within this pH range. It lacks an amide group.

References

Devlin, T. M. *Textbook of Biochemistry with Clinical Correlations* (3rd ed.). New York: Wiley-Liss, 1992. Pp. 6–13, 34–41, 1039–1056.

Mathews, C. K., and van Holde, K. E. *Biochemistry*. Redwood City, Calif.: Benjamin/Cummings, 1990. Pp. 41–50.

Murray, R. K., Granner, D. K., Mayes, P. A., and Rodwell, V. W. *Harper's Biochemistry* (22nd ed.). Norwalk, Conn.: Appleton & Lange, 1990. Pp. 13–19.

Stryer, L. *Biochemistry* (3rd ed.). New York: Freeman, 1988. Pp. 41–42.

Energetics

The three fundamental thermodynamic variables are enthalpy (H), entropy (S), and free energy (G).

In biochemistry the standard state is defined as pH 7.0, 25°C (298°K), all solutes at 1 molar concentration, and all gases at 1 atm pressure. A superscript zero prime is used to denote standard state conditions in biochemistry, as in $G^{0\prime}$, $H^{0\prime}$, and $S^{0\prime}$. The inorganic standard states, designated by a superscript zero, differ in that the standard pH is 0.0. The actual conditions are indicated by G, H, and S.

Enthalpy

Enthalpy, H, is defined as the heat content of a physical object or body; it is the sum of the internal energy (E) plus the pressure-volume product (PV):

$$H = E + PV \tag{4-1}$$

The **standard enthalpy of formation** of a chemical compound A, $\Delta H_F^{0\prime}$ (A), is the enthalpy change, or increase of heat content, due to the reaction that generates one mole of A from its constituent elements in their standard states, e.g., O_2, N_2, C (graphite), S, H_2, or whatever. The enthalpy of an element is arbitrarily taken to be zero. For the reaction below, the standard enthalpy change, $\Delta H^{0\prime}$, is calculated by subtracting the standard enthalpies of formation of the reactants from those of the products:

$$A + B \longrightarrow C + D$$
$$\Delta H^{0\prime} = \Delta H_F^{0\prime}(C) + \Delta H_F^{0\prime}(D) - \Delta H_F^{0\prime}(A) - \Delta H_F^{0\prime}(B) \tag{4-2}$$

Entropy

Entropy, S, is defined as the degree of randomness or disorder of a system. Carbon dioxide gas inside a bottle has a high entropy, because the molecules can move freely in any direction. The glass of the bottle, on the other hand, has a rigid structure that constrains freedom of molecular movement within that structure; hence, it has a low entropy. The entropy of a pure, crystalline solid at 0°K (absolute zero) is zero. The entropy increases as the substance goes from solid to liquid phase and from liquid to gas.

The **second law of thermodynamics** predicts that the entropy of a closed system will always increase. A closed system is one where no energy is being put in from an outside source. Unfortunately, the entropy change, ΔS, cannot be measured directly. Instead, ΔS is calculated from ΔH, ΔG, and ΔT, where ΔH is the enthalpy change, ΔG the free energy change, and ΔT the temperature change.

If $\Delta S < 0$, the randomness of the system declines; that is, the products have more order than the reactants. If $\Delta S > 0$, the products have a greater degree of disorder than the reactants.

Free Energy

The **free energy,** G, is the maximum usable work that can be obtained from a system at constant pressure, temperature, and volume. The energy spent to maintain randomness, TS (where T is measured in degrees Kelvin), cannot be harnessed to perform work. Hence, at constant temperature:

$$G = H - TS \tag{4-3}$$

$$\Delta G = \Delta H - T\Delta S \tag{4-4}$$

The **standard free energy of formation** of A, $\Delta G_F^{0\prime}(A)$, is defined as the change in free energy during the synthesis of 1 mole of A from its constituent elements under standard conditions.

The **standard free energy change** of a reaction, $\Delta G^{0\prime}$, can be calculated by subtracting the $\Delta G_F^{0\prime}$ values of the reactants from those of the products:

$$\Delta G^{0\prime} = \Delta G_F^{0\prime}(C) + \Delta G_F^{0\prime}(D) - \Delta G_F^{0\prime}(A) - \Delta G_F^{0\prime}(B) \tag{4-5}$$

The standard free energy of reaction, $\Delta G^{0\prime}$, can also be calculated from the **equilibrium constant** for the reaction, K_{eq}. The actual free energy change (ΔG) in calories per mole is related to $\Delta G^{0\prime}$:
where

$$K = \frac{[C] \times [D]}{[A] \times [B]}$$

$$\Delta G = \Delta G^{0\prime} + 4.57 \, T \log K \tag{4-6}$$

and T is in degrees Kelvin. The factor of 4.57 relates to the conversion from natural (ln) to common logarithms.

In the standard state, K equals 1.0, since by definition each reactant and product is at 1 molar concentration, and ΔG equals $\Delta G^{0\prime}$.

At equilibrium, K equals K_{eq} and ΔG equals zero (a system at equilibrium produces no usable work). Substituting into Equation 4-6 and rearranging:

$$\Delta G^{0\prime} = -4.57\ T \log K_{eq} \tag{4-7}$$

Having calculated $\Delta G^{0\prime}$ in calories per mole from K_{eq} using Equation 4-7, one can now use Equation 4-6 to calculate ΔG for any actual temperature and reactant concentrations.

If $\Delta G < 0$, then the reaction can proceed spontaneously. Such a reaction is **exergonic,** because it releases energy that can perform work. Only if properly harnessed, however, will this energy perform actual work.

If $\Delta G > 0$, then the forward reaction will not proceed spontaneously. Outside energy must be added to drive the reaction forward from A and B to the products C and D. Such a reaction is **endergonic,** or energy-requiring. The reverse reaction—i.e., from C and D to A and B—will have a negative value of ΔG and will tend to proceed spontaneously.

If ΔG equals zero, then the reaction is at equilibrium, i.e., the rate of the forward reaction is equal to the rate of the reverse reaction. The concentrations of the reactants and products at this point define the equilibrium constant, K_{eq}, for the reaction. (These concentrations need not be 1 molar, so the equilibrium state is different from the standard state.)

The actual conditions of most reactions—i.e., temperature, pressure, or concentration—can be adjusted to make ΔG positive, zero, or negative. However, it must be kept in mind that ΔG *gives no information about the reaction rate; it only indicates the direction of spontaneous reaction.* Its negative value may predict a forward reaction, but the reaction may proceed very slowly.

Living beings require a constant energy supply from their surroundings to maintain their high degree of internal order. Most individual reactions inside a cell reach equilibrium only when the cell dies. During life, these reactions accomplish rapid metabolic interconversions and operate nowhere near their equilibrium points.

High- and Low-Energy Phosphate Compounds

Bond energy, to an inorganic chemist, is the energy needed to break a bond. When biochemists refer to high- and low-energy phosphate bonds, however, they are talking about the $\Delta G^{0\prime}$ for the reaction that hydrolyzes the bond, instead of the energy required to break the bond itself.

Glycerol-l-phosphate has a low-energy phosphate bond, because $\Delta G^{0\prime}$ for phosphate hydrolysis is relatively low (-2.3 kcal/mole).

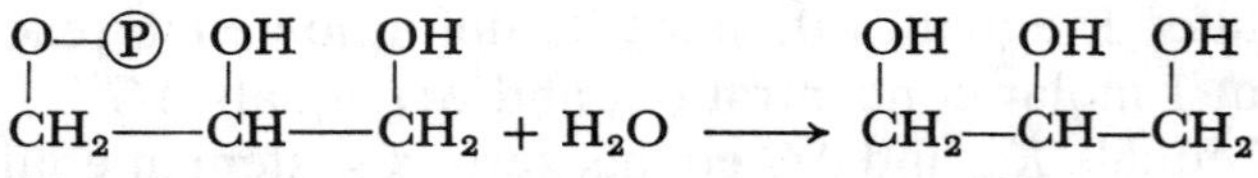

$$\begin{matrix} \text{O}-\text{\textcircled{P}} & \text{OH} & \text{OH} \\ | & | & | \\ \text{CH}_2-\text{CH}-\text{CH}_2 + \text{H}_2\text{O} \longrightarrow & & \end{matrix} \quad \begin{matrix} \text{OH} & \text{OH} & \text{OH} \\ | & | & | \\ \text{CH}_2-\text{CH}-\text{CH}_2 \end{matrix}$$

Glycerol-l-P   Glycerol

Adenosine triphosphate (ATP) is the principal intracellular energy currency. It has an intermediate value of $\Delta G^{0\prime}$ for its hydrolysis (-7.3 kcal/mole), but it is often called a high-energy phosphate compound. The two phosphoanhydride bonds within its triphosphate group account for its high energy on hydrolysis (-7.3 kcal/mole) in either of these reactions:

$$\text{ATP} + \text{H}_2\text{O} \rightleftharpoons \text{ADP} + \text{P}_i + \text{H}^+$$
$$\text{ATP} + \text{H}_2\text{O} \rightleftharpoons \text{AMP} + \text{PP}_i + \text{H}^+$$

ADP (adenosine diphosphate) has a lower free energy of hydrolysis than does ATP (-6.6 kcal/mole). AMP (adenosine monophosphate) has a much lower free energy of hydrolysis (-3.4 kcal/mole) and is therefore classified as a low-energy phosphate compound.

Pyrophosphate produced from hydrolysis of nucleoside triphosphates to nucleoside monophosphates can be further hydrolyzed to phosphate as shown below:

$$\text{PP}_i + \text{H}_2\text{O} \rightleftharpoons 2\text{P}_i \qquad \Delta G^{0\prime} = -4.6 \text{ kcal/mole}$$

Creatinine phosphate, the principal energy reservoir in muscle, has a phosphate bond of higher energy than that of ATP, with a $\Delta G^{0\prime}$ of -10.3 kcal/mole. In cells, the hydrolysis of such high-energy phosphate compounds is coupled to the phosphorylation of ADP to ATP. As shown below, creatine phosphate donates its phosphate to ADP to create ATP:

Creatine phosphate + $\text{H}_2\text{O} \longrightarrow$ creatine + P_i $\Delta G^{0\prime} = -10.3$ kcal/mole

ADP + $\text{P}_i \longrightarrow$ ATP + H_2O $\Delta G^{0\prime} = +7.3$ kcal/mole

---

Net: Creatine phosphate + ADP $\longrightarrow$ creatine + ATP $\Delta G^{0\prime} = -3.0$ kcal/mole

Other major high-energy phosphate compounds include phosphoenolpyruvate and 1,3-diphosphoglycerate. The hydrolysis of phosphate from each is coupled to ATP formation in glycolysis.

In the cell, endergonic reactions can be driven by coupling them to an exergonic reaction, such as ATP hydrolysis. The phosphorylation of glucose with P_i to glucose-6-phosphate, for instance, has a positive $\Delta G^{0\prime}$ of 3.3 kcal/mole. ATP donates the P_i to glucose and simultaneously supplies the energy to drive this reaction toward glucose-6-phosphate formation:

Glucose + $P_i \longrightarrow$ glucose-6-P + H_2O	$\Delta G^{0'} = +3.3$ kcal/mole
ATP + $H_2O \longrightarrow$ ADP + P_i	$\Delta G^{0'} = -7.3$ kcal/mole
Net: Glucose + ATP $\longrightarrow$ glucose-6-P + ADP	$\Delta G^{0'} = -4.0$ kcal/mole

Oxidation-Reduction Reactions

Oxidation is defined as loss of electrons, while **reduction** is a gain of electrons; Fe^{+2}, for example, is more reduced than Fe^{+3}, whereas Cu^{+2}, on the other hand, is more oxidized than Cu^+.

The **oxidation state** of a carbon atom depends on the electronegativity of the atoms bound to it. The carbon atom of methane (CH_4) represents the state of greatest reduction, because carbon and hydrogen equally share electrons. The carbon atom of methanol (CH_3OH) is more oxidized, because the hydroxyl oxygen is more electronegative than carbon. The C—OH bond is mildly polar; the carbon has a slight positive charge and has partially given up an electron to oxygen. The carbon atom of formaldehyde (HCHO) is even more oxidized; the carbonyl bond is more polarized than the C—OH bond of methanol. The carbon atom of formic acid (HCOOH) is more oxidized than that of formaldehyde. Carbon dioxide (O=C=O) has, in turn, a more oxidized carbon atom than does formic acid.

The **reduction potential,** $E^{0'}$, is the electrical potential (E) in volts (V) measured during the reduction reaction under standard conditions. The H_2:H^+ electrode is used as the reference or zero potential:

$$2H^+ + 2e^- \rightarrow H_2 \qquad E^{0'} = 0.00 \text{ V}$$

A compound that can be oxidized more readily than H_2 will have a negative $E^{0'}$. The **oxidation potential** of a reaction has the same absolute value as the reduction potential but the opposite sign.

An **oxidation-reduction reaction** involves a transfer of electrons among its reactants. Each overall oxidation-reduction reaction consists of two **half-reactions.** The mitochondrial electron-transport system, for example, couples the oxidation of NADH to NAD^+ and H^- (thus hydride ion represents $H^+ + 2e^-$) with the formation of water from $2H^+$, $\frac{1}{2}O_2$, and $2e^-$. The standard reduction potential of NAD^+ to NADH is -0.32 V. Therefore, the oxidation of NADH has an $E^{0'}$ of $+0.32$ V. The standard reduction potential for water formation from O_2, H^+, and e^- is $+0.816$ V. The difference between the standard reduction potentials, $\Delta E^{0'}$, is the **net reaction potential** under standard conditions, i.e., 25°C, all solutes at 1 molar concentration, and all gases at 1 atm pressure. To calculate $\Delta E^{0\prime}$, the oxidation potential of the oxidation half-reaction is added to the reduction potential of the reduction half-reaction:

$NADH \longrightarrow NAD^+ + H^+ + 2e^-$	$E = -E^{0\prime} = +0.320$ V
$\frac{1}{2}O_2 + 2H^+ + 2e^- \longrightarrow H_2O$	$E = E^{0\prime} = +0.816$ V
Net: $NADH + \frac{1}{2}O_2 + H^+ \longrightarrow NAD^+ + H_2O$	$\Delta E^{0\prime} = 1.14$ V

When ΔE, the reaction potential under actual (not necessarily standard) conditions, is positive, a reaction can proceed spontaneously. When ΔE equals zero, the reaction is at equilibrium.

Knowing $\Delta E^{0\prime}$ and n, the number of electrons transferred in the reaction, one can calculate $\Delta G^{0\prime}$ in kilocalories per mole:

$$\Delta G^{0\prime} = -23.1 \, (n) \, (\Delta E^{0\prime}) \tag{4-8}$$

Activation Energy

Activation energy, E_a, is the free energy needed to convert the reactants, such as the substrates of an enzymatic reaction, into their reactive states; once converted to the reactive or transition state, they are rapidly converted to products. The **reaction rate** is proportional to the quantity of reactants in the transition state. The higher the E_a, the slower the rate of reaction, because fewer of the reactant molecules will possess enough kinetic energy to become transformed into their reactive states.

Fig. 4-1   Effect of catalysts on activation energy (E_a).

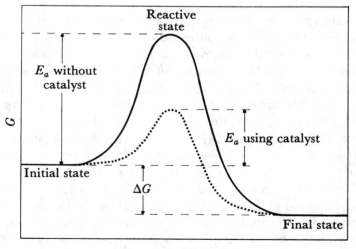

Progress of Reaction

One method for increasing the rate of an inorganic reaction is to raise the temperature, thereby promoting more of the reactants to the transition state. Another method is to add inorganic catalysts, such as platinum, which augment both the forward and reverse reaction rates by lowering the E_a barrier, as shown in Figure 4-1.

Since mammals cannot significantly raise their body temperature, they must rely on enzymes as catalysts to lower the E_a values for many reactions. Without enzymes, life could not continue.

Problems

Problem 1

The pyruvate kinase reaction is shown below. The $\Delta G^{0\prime}$ values for the hydrolysis of phosphate from phosphoenolpyruvate (PEP) and ATP are -14.8 and -7.3 kcal/mole, respectively. Calculate $\Delta G^{0\prime}$ for the reaction. Is it endergonic or exergonic?

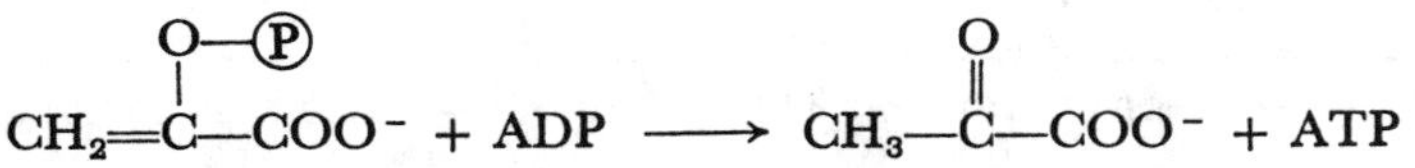

$$CH_2{=}\overset{\overset{\displaystyle O{-}\textcircled{P}}{|}}{C}{-}COO^- + ADP \longrightarrow CH_3{-}\overset{\overset{\displaystyle O}{\|}}{C}{-}COO^- + ATP$$

PEP   Pyruvate

(handwritten:)
-14.8
$+7.3$
$\overline{-7.5}$ (<0)

Problem 2

Alcoholics tolerate alcohol better than nondrinkers because their liver has more alcohol dehydrogenase to oxidize ethanol to acetaldehyde, as shown below. Choose the single best answer.

$$NAD^+ + H^+ + 2e^- \longrightarrow NADH \qquad E = E^{0\prime} = -0.320\ V$$
$$CH_3CH_2OH \longrightarrow CH_3CHO + 2H^+ + 2e^- \qquad E = -E^{0\prime} = +0.197\ V$$

Net: $CH_3CH_2OH + NAD^+ \longrightarrow CH_3CHO + NADH + H^+$

A. $\Delta E^{0\prime} = -0.517\ V$.
B. $\Delta G^{0\prime} = +2.84$ kcal/mole.
C. $\Delta G^{0\prime} = +5.68$ kcal/mole.
D. $\Delta G^{0\prime}$ cannot be determined from the data given.

(handwritten:)
-0.123
$\Delta G = -23.1 (n)(\Delta E^{0\prime})$
$-\Delta G = -23.1 \times 2 \times (-0.123)$
$\Delta G = +5.68$

Problem 3

$\Delta G^{0\prime}$ for reactions X and Y are -10 and -5 kcal/mole, respectively. Choose the best single answer about these reactions under standard conditions:

A. The rate of reaction X will exceed that of reaction Y.
B. E_a of reaction Y will exceed that of reaction X.
C. Neither will proceed without additional free energy.
D. Reaction rates cannot be deduced from these data.

Problem 4
Which of these reactions could not be driven by coupling them to the hydrolysis of ATP to ADP + P$_i$?

A. Creatine + P$_i$ → creatine phosphate + H$_2$O
$$\Delta G^{0\prime} = +10.3 \text{ kcal/mole}$$

B. Glycerol + P$_i$ → glycerol-1-phosphate + H$_2$O
$$\Delta G^{0\prime} = +2.2 \text{ kcal/mole}$$

C. Glucose + P$_i$ → glucose-6-phosphate
$$\Delta G^{0\prime} = +3.3 \text{ kcal/mole}$$

D. Fructose-6-phosphate + P$_i$ → fructose-1,6-biphosphate + H$_2$O
$$\Delta G^{0\prime} = +4.0 \text{ kcal/mole}$$

Answers

1. This reaction is exergonic, as shown below.

PEP + H$_2$O $\longrightarrow$ pyruvate + P$_i$	$\Delta G^{0\prime} = -14.8$ kcal/mole
ADP + P$_i$ $\longrightarrow$ ATP + H$_2$O	$\Delta G^{0\prime} = +7.3$ kcal/mole
Net: PEP + ADP $\longrightarrow$ pyruvate + ATP	$\Delta G^{0\prime} = -7.5$ kcal/mole

2. C. $\Delta E^{0\prime} = -0.320 - (-0.197) = -0.123$ V. $\Delta G^{0\prime} = -23.1(n) (\Delta E^{0\prime})$. Substituting $n = 2$ and $\Delta E^{0\prime} = -0.123$, $\Delta G^{0\prime} = +5.68$ kcal/mole.
3. D. From the data supplied on $\Delta G^{0\prime}$, one cannot deduce reaction rates or activation energies.
4. A. ATP hydrolysis to ADP + P$_i$ liberates 7.3 kcal/mole, enough to drive reactions B, C, and D.

References

Devlin, T. M. *Textbook of Biochemistry with Clinical Correlations* (3rd ed.). New York: Wiley-Liss, 1992. Pp. 237–247, 270–273.

Mathews, C. K., and van Holde, K. E. *Biochemistry*. Redwood City, Calif.: Benjamin/Cummings, 1990. Pp. 59–87.

Murray, R. K., Granner, D. K., Mayes, P. A., and Rodwell, V. W. *Harper's Biochemistry* (22nd ed.). Norwalk, Conn.: Appleton & Lange, 1990. Pp. 99–104.

Stryer, L., *Biochemistry* (3rd ed.). New York: Freeman, 1988. Pp. 315–320, 395–401.

Enzymes

Enzymes can best be defined as catalysts with a high degree of specificity for a certain substrate or class of substrates. As a catalyst, they lower the activation energy of chemical reactions. Like other catalysts, enzymes in very low concentrations enhance both the forward and reverse reaction rates without themselves being consumed. Unlike the metallic and inorganic catalysts, however, each enzyme can act on only one substrate or on a family of structurally similar substrates.

Virtually all enzymes contain a polypeptide or protein. Some also have nonprotein prosthetic groups, such as heme, heavy metals, or coenzymes. Coenzymes are nonprotein, organic molecules that assist enzymes in transferring certain groups (see Chap. 6).

In rare cases RNA can serve as enzymes (e.g., RNA can function as RNA polymerase and ribonuclease).

Substrates bind noncovalently to the substrate-binding site of enzymes. The substrate-binding site may or may not be located at the active or catalytic site of the enzyme. Prosthetic groups are also attached to the active site. Active sites are found in the crevices and fissures of enzymes. The amino acids at the active site serve as the catalysts.

Enzyme Nomenclature

In general, the name of an enzyme consists of two parts: first, the name of the substrate, or occasionally that of the product, is stated; the second portion of the name describes the type of reaction.

The International Union of Biochemists (IUB) has assigned a recommended name for each enzyme to replace its historical name or names. Some texts have adopted these recommended names, whereas others continue to use the former names. This text will generally use the recommended names. The IUB has also assigned a systematic name for each enzyme, which is often too long and cumbersome to be adopted for general usage.

Enzyme Types

Aldolase: Cleaves a carbon-carbon bond to create an aldehyde group.

Carboxylase: Adds CO_2 or HCO_3^- to its substrate to form a carboxyl group.

Decarboxylase: Cleaves a carboxyl group, e.g., from α-keto acids, liberating it as CO_2.

Dehydrogenase: Removes hydrogen atoms from its substrate.

Esterase: Hydrolyzes ester linkages to form an acid and an alcohol.

Hydratase: Adds water to a carbon-carbon double bond without breaking the bond or, conversely, removes water to create a double bond.

Hydrolase: Adds water to break a bond (hydrolysis). The suffix "ase" alone often denotes a hydrolase; e.g., sucrase hydrolyzes sucrose.

Hydroxylase: Incorporates an oxygen atom from O_2 into its substrate to create a hydroxyl group.

Isomerase: Converts between *cis* and *trans* isomers, D and L isomers, or aldose and ketose.

Kinase: Transfers a phosphate group from a high-energy phosphate compound, such as ATP, to its substrate (in contrast, a phosphorylase adds inorganic phosphate, P_i, to its substrate).

Ligase: Joins two molecules together using the energy released from hydrolyzing a pyrophosphate bond of a high-energy phosphate compound; also called synthetase.

Lyase: Breaks C—C, C—O, or C—N bonds, other than by hydrolysis or by oxidation-reduction.

Mutase: Shifts the position of a group, e.g., a methyl group, within a single molecule.

Oxidase: Adds O_2 to hydrogen atoms removed from the substrate (which is thereby oxidized) to generate H_2O_2, H_2O, or O_2^- (superoxide).

Oxygenase: Incorporates molecular O_2 into its substrates.

Peptidase: Hydrolyzes peptide bonds to yield free amino acids and peptides.

Phosphatase: Hydrolyzes substrates, such as phosphoric esters, to liberate inorganic phosphate (P_i; at physiologic pH, a mixture of HPO_4^{-2} and $H_2PO_4^-$).

Phosphorylase: Adds inorganic phosphate (P_i) to split a bond (phosphorolysis).

Polymerase: Synthesizes polymers such as DNA.

Protease: Cleaves peptide bonds.

Reductase: Catalyzes the reduction of its substrate, i.e., adds hydrogen atoms.

Sulfatase: Hydrolyzes substrates, such as sulfuric acid esters, to liberate sulfate.

Synthase: Joins two molecules together without hydrolyzing a pyrophosphate bond (in contrast, ligase or synthetase requires the hydrolysis of such a bond).

Synthetase: Same as ligase.

Transaminase: Transfers amino groups from an amino acid to a keto acid (also known as aminotransferase).

Transferase: Transfers groups containing carbon, nitrogen, phosphorus, or sulfur for phosphotransferases or methyl groups for methyltransferases, from one molecule to another without concomitant oxidation or reduction.

Enzyme Kinetics

Enzymes catalyze reactions by bringing substrates together in the best orientation to facilitate reaction. They reduce the free energy required to reach the transition state. Enzyme (E) combines with substrate (S) to form an enzyme-substrate complex (ES). Initially it was thought that substrates resembled keys fitting into a lock, the active site of the enzyme. It now appears that the interaction between enzymes and substrates can induce a fit between the two by changing the conformation of the active site.

Factors that alter enzyme-substrate binding include pH and temperature. Because of the charge that may exist on ionizable R groups near the active site at high or low pH values, ionic bonds or electrostatic attractions or repulsions may be present that can enhance, diminish, or prevent substrate binding. Hence, each enzyme has an **optimal pH value** at which it has maximal activity. This optimal pH is often near the pH of the tissue that contains the enzyme; for example, pepsin secreted into the stomach has an optimal pH of about 2.0 (gastric pH is 2.0 to 3.0); pancreatic α-amylase, when secreted into the duodenum, has an optimal pH of about 7.0 (intestinal pH is 6.0 to 7.0); and alkaline phosphatase in bone has an optimal pH of 9.0 to 10.0 (bone pH is above 7.4).

Raising the temperature increases the reaction rate (provided the higher temperature does not denature the enzyme), because it increases the kinetic energy of the molecules, thus allowing more frequent collisions between the enzyme and the substrate.

Reaction Order

For the reaction below, let k be the **rate constant,** R the **reaction order,** and V the **rate:**

$$S \xrightarrow{\quad k \quad} P$$
$$V = k[S]^R \tag{5-1}$$

For a **zero-order reaction** with respect to $[S]$, the rate is constant regardless of $[S]$:

$$V = k[S]^0 = k \qquad \text{(zero order)} \tag{5-2}$$

In **first-order reactions** with respect to $[S]$, the rate is proportional to $[S]$, while in **second-order reactions,** the rate is proportional to $[S]^2$:

$$V = k[S] \qquad \text{(first order)} \tag{5-3}$$
$$V = k[S]^2 \qquad \text{(second order)} \tag{5-4}$$

Michaelis-Menten Equation

The enzyme-substrate complex (ES) has two fates: S can be converted to P (product) or ES can dissociate back to $E + S$, as represented below:

$$E + S \underset{k_2}{\overset{k_1}{\rightleftharpoons}} ES \xrightarrow{\quad k_3 \quad} P + E \tag{5-5}$$

Michaelis and Menten reasoned that by measuring the initial reaction velocity (V) rather than later rates (when much of S had been converted to P), they could assume that $[S]$ remained at $[S]_0$, the initial substrate concentration, and that $[P]$ was zero. They then set up mathematical expressions for the rate of all three reactions in Equation 5-5, defining V_{max} as the velocity extrapolated to infinitely high $[S]$ and K_m as $(k_2 + k_3)/k_1$. If $[ES]$ remains constant (the steady-state assumption), then these rate expressions become:

$$V = \frac{V_{max}[S]_0}{K_m + [S]_0} \tag{5-6}$$

Equation 5-6 states the **Michaelis-Menten equation,** where K_m, the **Michaelis constant,** represents the substrate concentration at which the reaction rate is one-half V_{max}. It is *not* an equilibrium constant. Furthermore, since k_1, k_2, and k_3 cannot be empirically measured, K_m cannot be calculated from them; instead, K_m must be determined experimentally after measuring V_{max} and V at various substrate concentrations.

With certain enzymes, the dissociation of ES to an enzyme-product complex (EP) is the rate-limiting step. In such circumstances, K_m becomes the dissociation content for the enzyme-substrate complex (ES).

Enzymes with several substrates will possess a different K_m for each substrate. Both K_m and V_{max} for each enzyme will vary with changes in pH and temperature.

By taking the reciprocal of the Michaelis-Menten equation, one gets the **Lineweaver-Burk equation**:

$$\frac{1}{V} = \frac{K_m}{V_{max}[S]_0} + \frac{1}{V_{max}} \tag{5-7}$$

The double-reciprocal plot of $1/V$ versus $1/[S]_0$ is quite useful, because its y-intercept is $1/V_{max}$ while its x-intercept is $-1/K_m$, as illustrated in Figure 5-1.

Enzyme Inhibitors

There are two classes of enzyme inhibitors: reversible and irreversible. **Irreversible inhibitors** bind covalently to enzymes and dissociate very slowly, as indicated by the thickness of the arrows below:

$$E + I \rightleftharpoons EI$$

Toxic organophosphorous compounds, such as certain insecticides and diisopropyl fluorophosphate (DFP), act as irreversible inhibitors by binding to the active site of human acetylcholinesterase, which leads to a toxic accumulation of acetylcholine.

Fig. 5-1 Lineweaver-Burk plot.

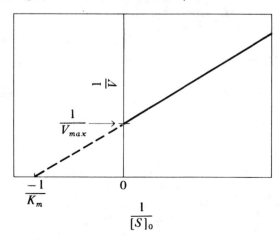

Cyanide and sulfide bind to the iron atom of cytochrome oxidase, causing irreversible inhibition. Not all irreversible inhibitors, however, bind to the active site of enzymes.

Reversible inhibitors bind noncovalently to enzymes through hydrogen bonds or salt bonds. Unlike the case of irreversible inhibitors, the reactions affected obey Michaelis-Menten kinetics.

Reversible enzyme inhibitors may have either of two modes of action: competitive inhibition or noncompetitive inhibition.

Competitive inhibitors bind to the active sites of enzymes and they compete with the substrate for enzyme binding. Although competitive inhibitors usually resemble the substrate in structure, this is not always the case. Competitive inhibition is reversible, since high substrate concentrations will overcome the effect of the inhibitor. In such reactions, the V_{max} will be reached, but K_m increases with increasing inhibitor concentration because the inhibitor reduces substrate binding to the catalytic site. Thus, the degree of inhibition depends upon the ratio $[I]/[S]$, where $[I]$ represents the inhibitor concentration. Figure 5-2 shows the characteristic increase in K_m with no change in V_{max} in the presence of a competitive inhibitor.

Fig. 5-2   Effect of competitive inhibitor concentration, $[I]$, on K_m and V_{max} of a hypothetical enzyme (see Problem 3).

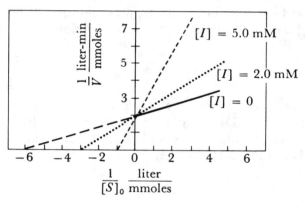

Fig. 5-3   Effect of noncompetitive inhibitor concentration, $[I]$, on K_m and V_{max} of a hypothetical enzyme (see Problem 4).

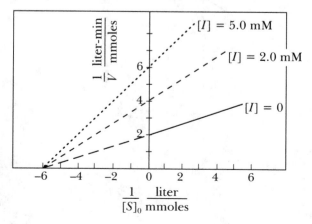

Noncompetitive inhibitors bind to enzymes in areas other than the active site. Unlike competitive inhibitors, they do not resemble the substrate. The degree of inhibition depends upon $[I]$; raising $[S]$ will not overcome the inhibition. Hence, in contrast to competitive inhibition, V_{max} decreases. As is the case with competitive inhibition, however, K_m usually increases (Fig. 5-3), but it occasionally remains the same.

Heavy metals such as Hg^{+2} and Pb^{+2} bind to the sulfhydryl groups in enzymes and inhibit noncompetitively.

Irreversible inhibitors bind covalently to enzymes. They are neither competitive nor noncompetitive.

Regulation of Enzyme Activity

Four mechanisms are commonly used to control enzyme activity:

1. Proteolytic activation. In this mechanism an inactive enzyme is irreversibly converted into the active form. Digestive enzymes, for example, are secreted as proenzymes and later converted to active enzymes. Coagulation factors are often synthesized as inactive proteins. Insulin is secreted as proinsulin, which is subsequently converted to insulin.
2. Control proteins may inhibit or stimulate enzymes. Calmodulin, for example, stimulates a wide variety of enzymes.
3. Reversible covalent modification. The phosphorylation of glycogen phosphorylase activates this enzyme.
4. Allosteric control. Allosteric control occurs via noncovalent binding.

The last three mechanisms involve reversible rather than irreversible enzyme activation.

The word **allosteric** means "another site." By definition, an **allosteric enzyme** has a regulatory site, which differs from its catalytic site, that binds allosteric effectors, also called modulators or modifiers. Virtually all allosteric enzymes have multiple polypeptide subunits. Usually, noncovalent bonds are formed between the effector and the enzyme.

Positive or **stimulatory effectors** enhance substrate binding, whereas **negative effectors** reduce substrate binding. Upon binding to the regulatory site, effectors change the quaternary structure of the allosteric enzyme, thereby altering the substrate-binding properties of the catalytic site.

Unlike a competitive inhibitor, an allosteric modifier need not resemble its enzyme's substrate, although it may do so occasionally.

Allosteric enzymes are often strategically placed at the first, or **committed step,** of a long metabolic pathway. The final endproduct of the pathway may then act as a negative modulator for that enzyme; this is termed **endproduct** or **feedback inhibition.** Such is the case for aspartate transcarbamoylase (ATCase), a transferase that controls the rate of pyrimidine synthesis. ATCase adds carbamoyl phosphate

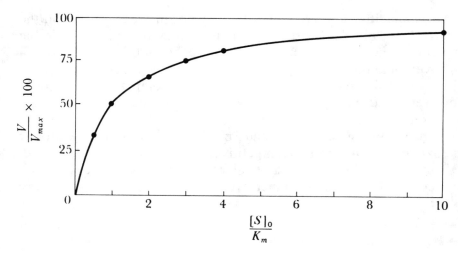

Fig. 5-4   Rectangular-hyperbolic plot that is characteristic of the reactions of nonallosteric enzymes (compare with Problem 2).

to aspartate to yield N-carbamoylaspartate, an intermediate in pyrimidine biosynthesis. In bacteria, but not in mammals, this enzyme is allosterically regulated. Cytidine triphosphate (CTP), an endproduct of pyrimidine synthesis, acts as a negative allosteric modulator in bacteria. When the CTP level becomes high, the enzyme is inhibited, thus shutting off further synthesis. On the other hand, the purine nucleoside phosphates (such as ATP) allosterically stimulate ATCase, primarily by preventing the binding of CTP. Not all endproducts are negative modulators. Similarly not all negative modulators are endproducts.

Regardless of the presence or absence of its effectors, an allosteric enzyme does not obey Michaelis-Menten kinetics. When V versus $[S]_0$ is graphed for an allosteric enzyme reaction, the curve deviates from the rectangular hyperbola of a nonallosteric enzymatic reaction, shown in Figure 5-4. Most allosteric enzyme reactions display a sigmoidal curve, which indicates that the binding of one substrate facilitates the binding of additional substrate molecules to other active sites. This sigmoidal curve also appears for the oxygen-binding reaction of the nonenzymatic protein, hemoglobin (see Fig. 2-1).

Isozymes

Enzymes that contain multiple polypeptide subunits of two or more types and that exist in several different forms are termed **isozymes.** Lactate dehydrogenase (LDH), for example, is a tetramer of heart (H) and muscle (M) subunits. Five isozymes of LDH exist: H_4, H_3M, H_2M_2, HM_3, and M_4. These isozymes differ from one another with respect to the K_m and V_{max} values for the reaction with lactate.

Creatine phosphokinase (CPK) is a dimer consisting of M and brain (B) type subunits. Skeletal muscle contains the MM, cardiac muscle the MB, and brain the MM isozyme. CPK isozyme determination is invaluable in the diagnosis of acute myocardial infarction (heart attack); elevated MB isozyme is specific for myocardial damage.

Enzyme Activity

The international unit (IU) of enzyme activity is defined as the quantity of enzyme needed to transform 1.0 micromole of substrate to product per minute at 30°C and optimal pH. To determine the enzyme activity in a blood sample, for instance, one must measure the rate of the enzymatically catalyzed reaction with excess substrate present to ensure that the enzyme has been saturated and that the rate depends only on the enzyme's activity.

Problems

Problem 1
Estimate the reaction order with respect to $[S]$ for tracings A–D in Figure 5-5.

Problem 2
Use the Michaelis-Menten equation to complete the table below. Graph your results on Figure 5-6, and determine the order of the reaction (note that since K_m and V_{max} are constants, your curve will have the same slope as one of the V versus $[S]_0$).

$\dfrac{[S]_0}{K_m}$	$\dfrac{V}{V_{max}} \times 100$
$\frac{1}{2}$	33
1	. . .
2	. . .
3	. . .
10	. . .

Fig. 5-5 Graph of initial velocity (*V*) vs initial substrate concentration ($[S]_0$).

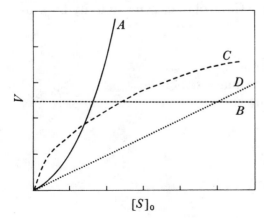

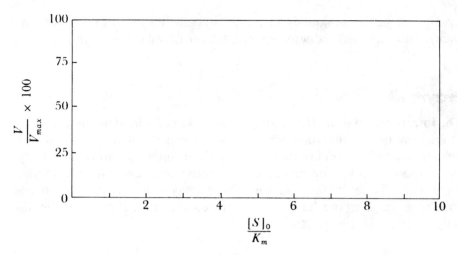

Fig. 5-6   Graph your results from Problem 2 here to determine the reaction order.

Problem 3
Calculate the K_m and V_{max} for the hypothetical enzyme with the competitive inhibitor in Figure 5-2, when $[I] = 0$, 2.0, and 5.0 mM.

Problem 4
Calculate K_m and V_{max} for the hypothetical enzyme with the noncompetitive inhibitor shown in Figure 5-3, when $[I] = 0$, 2.0, and 5.0 mM.

Problem 5
The K_m of hexokinase for glucose is 0.15 mM, whereas its K_m for fructose is 1.5 mM. Assume V_{max} is the same for both substrates. Choose the single best answer:

    A. At substrate concentration of 0.1 mM, hexokinase will phosphorylate fructose more rapidly than glucose.
    B. At substrate concentration above 5 mM, both reactions will proceed at similar rates.
    C. Competitive inhibitors of hexokinase would reduce K_m.
    D. Competitive inhibitors would reduce V_{max}.

Problem 6
Choose the single best statement. Control of enzyme activity by proteolytic activation

    A. Is irreversible.
    B. Involves feedback inhibition.
    C. Alters secondary but not primary structure.
    D. Does not occur outside the gut.

Answers

1. A.   Second-order.
   B.   Zero-order.

    C.   Mixed zero- and first-order.

    D.   First-order.

2. $\dfrac{V}{V_{max}} = \dfrac{[S]_0}{K_m + [S]_0} = \left(\dfrac{[S]_0}{K_m}\right) \cdot \left(\dfrac{1}{1 + [S]_0 / K_m}\right)$

Now substitute the $\frac{1}{2}$, 1, 2, 3, and 10 into the $[S]_0 / K_m$ expression.

$\dfrac{[S]_0}{K_m}$	$\dfrac{V}{V_{max}} \times 100$
$\frac{1}{2}$	33
1	50
2	67
3	75
10	91

3. For all inhibitor concentrations, $V_{max} = 0.50$ mM/liter-min.
   When $[I] = 0$, $K_m = 0.167$ mM.
   When $[I] = 2.0$ mM, $K_m = 0.333$ mM.
   When $[I] = 5.0$ mM, $K_m = 1.0$ mM.
4. $K_m = 0.167$ mM for all inhibitor concentrations.
   When $[I] = 0$, $V_{max} = 0.50$ mM/liter-min.
   When $[I] = 2.0$ mM, $V_{max} = 0.25$ mM/liter-min.
   When $[I] = 5.0$ mM, $V_{max} = 0.167$ mM/liter-min.
5. B.   At low substrate concentration, hexokinase phosphorylates glucose more rapidly than fructose. At high substrate concentrations relative to K_m, both reactions proceed at a similar rate. Competitive inhibitors increase K_m but do not alter V_{max}.
6. A.   Feedback inhibition applies to allosteric enzymes. Proinsulin and coagulation factors are activated outside the gut by proteolytic enzymes.

References

Devlin, T. M. *Textbook of Biochemistry with Clinical Correlations* (3rd ed.). New York: Wiley-Liss, 1992. Pp. 135–191.

Mathews, C. K., and van Holde, K. E. *Biochemistry*. Redwood City, Calif.: Benjamin/Cummings, 1990. Pp. 339–365, 373–403.

Murray, R. K., Granner, D. K., Mayes, P. A., and Rodwell, V. W. *Harper's Biochemistry* (22nd ed.). Norwalk, Conn.: Appleton & Lange, 1990. Pp. 58–97.

Stryer, L. *Biochemistry* (3rd ed.). New York: Freeman, 1988. Pp. 177–257.

Vitamins and Coenzymes

A **coenzyme** is a nonprotein organic molecule that binds to an enzyme to aid in the transfer of specific functional groups. Usually, it binds loosely and can be easily separated from its enzyme, but when it binds tightly, it is considered to be a **prosthetic group** of the enzyme.

A **cofactor** differs from a coenzyme only because it usually is a metallic ion rather than an organic molecule. Examples include Fe^{+2} in the cytochromes, Mg^{+2} for enzymes utilizing ATP, Zn^{+2} in lactate dehydrogenase, Mo^{+6} in xanthine oxidase, and Cu^{+2} in cytochrome oxidase.

In the process of transferring functional groups, both coenzymes and cofactors often change their own structure or valence and must be later returned to their original form.

For humans, a **vitamin** is an organic molecule required for certain metabolic functions that must be supplied in very small amounts (less than 50 mg/day) because we either cannot synthesize it or cannot synthesize enough to meet our needs. This definition excludes inorganic compounds, such as metals and minerals, and essential nutrients required in large amounts, such as amino acids, glucose, and triglycerides.

The **minimal daily requirement (MDR)** of a vitamin is the minimal oral intake necessary to prevent the symptoms and signs of vitamin deficiency from appearing. Meeting but not exceeding the MDR may still allow the biochemical abnormalities of vitamin deficiency to be present. Therefore, the MDR is not in itself an adequate vitamin intake.

The **recommended daily allowance (RDA)** of a vitamin represents an estimate of the adequate vitamin intake for the majority of healthy Americans in each age bracket. Meeting the RDA will guarantee an adequate vitamin supply for all but a small minority of Americans. An intake less than the RDA, however, does not necessarily cause a vitamin shortage.

Water-Soluble Vitamins and Coenzymes

The **water-soluble vitamins,** which include all B vitamins and vitamin C, act as coenzymes or coenzyme precursors (Table 6-1). Many serve as activated carriers, transferring groups ranging from electrons to

Table 6-1 Vitamins acting as coenzymes.*

Vitamin	Coenzyme	Metabolic role	Deficiency state
Thiamine	Thiamine pyrophosphate	Decarboxylation of α-keto acids and α-keto sugars Transketolase reaction	Wernicke-Korsakoff syndrome and beriberi
Riboflavin	FMN, FAD	Oxidative deamination of amino acids β-Oxidation of fatty acids Oxidative phosphorylation Purine catabolism	Angular stomatitis Glossitis Seborrheic dermatitis
Niacin	NADH, NADPH	Numerous oxidation-reduction reactions	Glossitis, pellagra
Pyridoxine	Pyridoxal phosphate	Transfer amino groups Decarboxylate amino acids	Peripheral neuropathy Glossitis Hypochromic anemia Niacin deficiency
Pantothenic acid	CoA	Transfer activated acyl groups	Not established
Biotin	Various carboxylations	Carry activated CO_2	Rare
Folic acid	THFA	Carry activated one-carbon groups	Macrocytic anemia Glossitis
B_{12}	Methylcobalamin Deoxyadenosyl-cobalamin	Carry activated one-carbon groups	Megaloblastic anemia Subacute combined degeneration of the spinal cord Peripheral neuropathy Glossitis

*The deficiency states listed are not necessarily related to the metabolic roles listed next to them.

FMN = flavin mononucleotide; FAD = flavin-adenine dinucleotide; NADH = reduced form of nicotinamide-adenine dinucleotide; NADPH = reduced form of nicotinamide-adenine dinucleotide phosphate; THFA = tetrahydrofolic acid; CoA = coenzyme A; CO_2 = carbon dioxide.

acyl groups or single carbon groups. The activated forms of these groups react readily, thereby assisting enzymes in catalysis.

Glossitis, the loss of tongue papillae, occurs in five of the B-vitamin deficiency states (folic acid, niacin, riboflavin, pyridoxine, and cyanocobalamin) as well as in iron deficiency.

In the following structural formulas, an asterisk (*) designates the reactive site, insofar as it has been determined.

Thiamine Pyrophosphate (TPP)

Thiamine pyrophosphate (TPP)

TPP is an ester of thiamine or vitamin B_1. It should not be confused with thymine, a pyrimidine.

The essential role of TPP is to serve as a coenzyme in the decarboxylation of α-keto acids and keto sugars. This mitochondrial reaction involves the hydrogenation and then the dehydrogenation of the α-keto group of pyruvate along with removal of its carboxyl group:

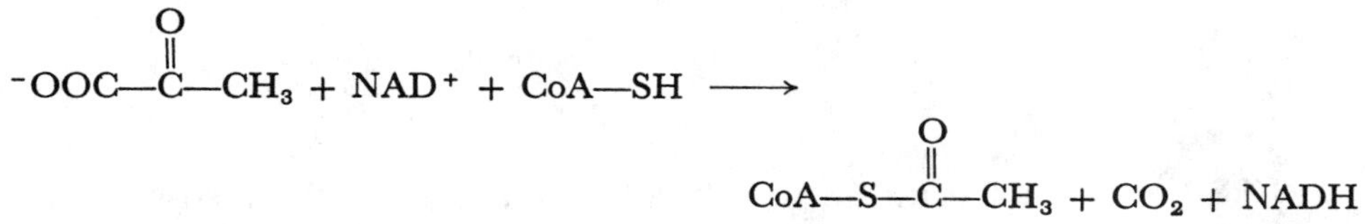

$$^-OOC-\overset{\overset{\displaystyle O}{\|}}{C}-CH_3 + NAD^+ + CoA-SH \longrightarrow$$

$$CoA-S-\overset{\overset{\displaystyle O}{\|}}{C}-CH_3 + CO_2 + NADH$$

This reaction occurs in four steps. Pyruvate dehydrogenase, with TPP as its coenzyme, catalyzes steps 1 and 2. Step 1 involves the initial decarboxylation of pyruvate, forming hydroxyethyl-TPP. In step 2, the disulfide group of lipoamide (oxidized lipoic acid bound to the amino side chain of lysine) oxidizes the hydroxyethyl group on TPP and transfers the resultant acetyl group to itself. In step 3, dihydrolipoyl transacetylase, with lipoamide as its coenzyme, transfers this acetyl group to CoA. In the fourth and final step, dihydrolipoyl dehydrogenase, with FAD as its coenzyme, removes the hydride ion from dihydrolipoamide and transfers it to NAD^+, regenerating lipoamide.

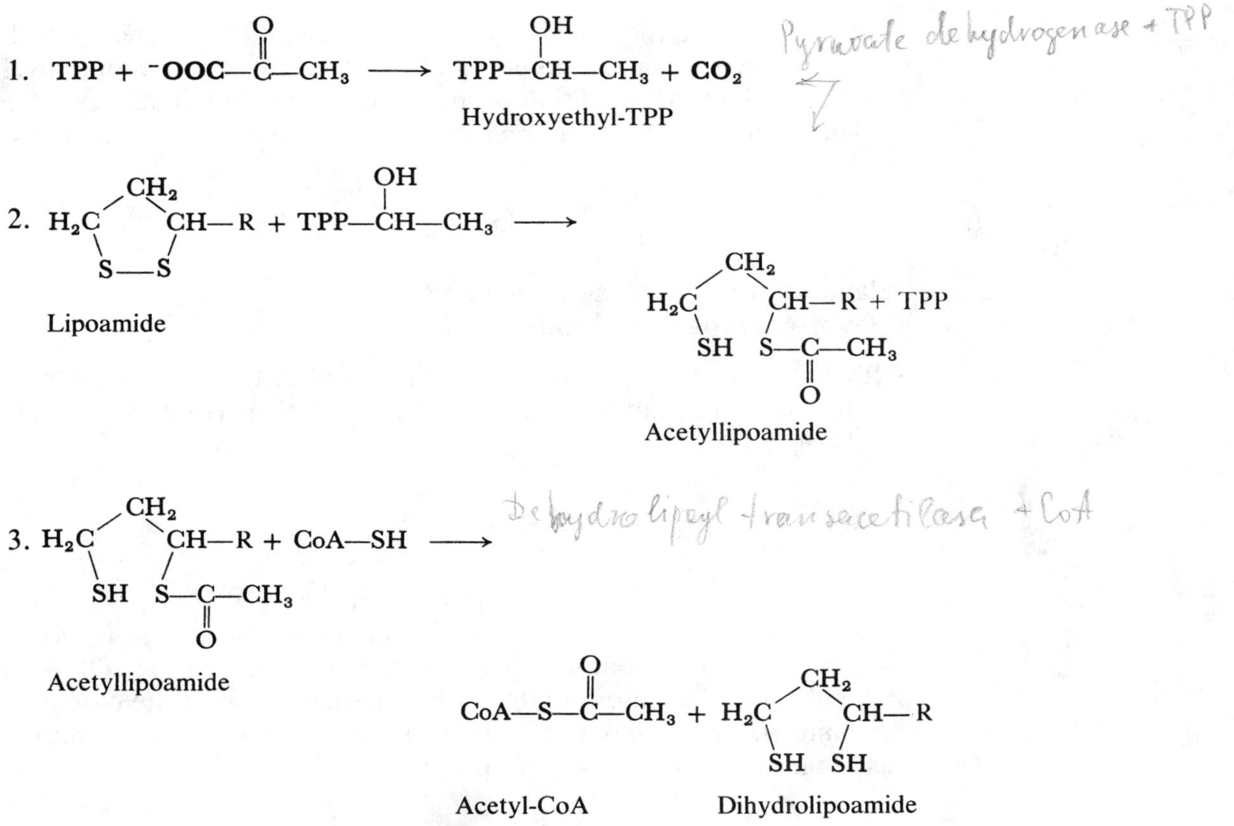

1. $TPP + {}^-OOC-\overset{\overset{\displaystyle O}{\|}}{C}-CH_3 \longrightarrow TPP-\overset{\overset{\displaystyle OH}{|}}{C}H-CH_3 + CO_2$ *Pyruvate dehydrogenase + TPP*

   Hydroxyethyl-TPP

2. Lipoamide + $TPP-\overset{\overset{\displaystyle OH}{|}}{C}H-CH_3 \longrightarrow$ Acetyllipoamide $+ TPP$

3. Acetyllipoamide $+ CoA-SH \longrightarrow$ *Dehydrolipoyl transacetilasa + CoA*

   $CoA-S-\overset{\overset{\displaystyle O}{\|}}{C}-CH_3 +$ Dihydrolipoamide

   Acetyl-CoA   Dihydrolipoamide

Dehydrolipoyl dehydrogenase + FAD

4. $\underset{\displaystyle \underset{\text{SH SH}}{|\quad|}}{\underset{\displaystyle H_2C \qquad CH{-}R}{\diagdown \quad \diagup}}\overset{\displaystyle CH_2}{}$ + NAD$^+$ $\longrightarrow$

Dihydrolipoamide

$\underset{\displaystyle \underset{\text{S}{-}\text{S}}{}}{\underset{\displaystyle H_2C \qquad CH{-}R}{\diagdown \quad \diagup}}\overset{\displaystyle CH_2}{}$ + NADH + H$^+$

Lipoamide

TPP is also used in the transketolase reaction of the hexose-monophosphate shunt to transfer a —(C=O)—CH$_2$OH group between two sugar phosphates.

Signs of thiamine deficiency appear in people whose caloric intake is disproportionately high compared to their thiamine intake. Such an imbalance occurs endemically in certain areas of Asia where people subsist largely on polished, milled rice (the processing of which removes the thiamine) and also in chronic alcoholics who eat little food (alcoholic beverages provide calories but not thiamine). Thus, the RDA for thiamine is stated in proportion to the caloric intake.

A moderate thiamine deficiency impairs carbohydrate metabolism in neurons, producing peripheral neuropathy ("dry" beriberi). Severe thiamine deficiency impairs carbohydrate metabolism in the heart and blood vessels, causing high-output congestive heart failure ("wet" beriberi).

Thiamine deficiency can cause acute confusion, ataxia, and ophthalmoplegia (Wernicke-Korsakoff syndrome), which may become irreversible unless the condition is quickly treated with thiamine.

The principal dietary sources of thiamine are meats, beans, peas, and grains.

Flavin Mononucleotide (FMN) and Flavin-Adenine Dinucleotide (FAD)

Riboflavin, or vitamin B$_2$, is phosphorylated in the intestine to generate FMN (riboflavin 5'-phosphate). FMN then reacts with ATP, yielding FAD:

FMN + ATP $\longrightarrow$ FAD + PP$_i$

where PP$_i$ indicates inorganic pyrophosphate.

Both FAD and FMN transfer hydrogen atoms and electrons, utilizing the two nitrogen atoms designated by asterisks in the following formula. They are used by flavin-linked enzymes as in the oxidative deamination of amino acids, the β-oxidation of fatty acids, purine catabolism, and oxidative phosphorylation.

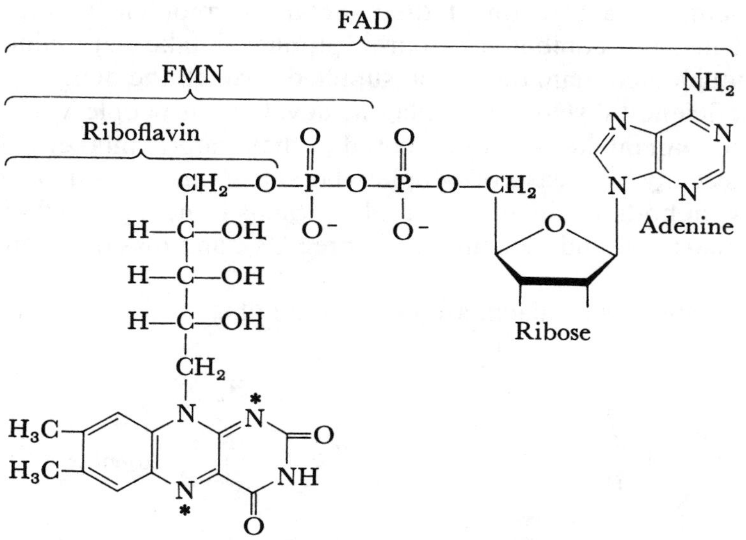

Flavin mononucleotide (FMN) and flavin-adenine dinucleotide (FAD)

Riboflavin deficiency leads to angular stomatitis (fissures at the angles of the mouth), localized seborrheic dermatitis of the face, vascular changes in the cornea, and a purple smooth tongue due to loss of tongue papillae.

The riboflavin requirement is proportional to the protein intake.

The principal dietary sources are dairy products and organ meats (liver and heart) but not muscle meats.

Nicotinamide-Adenine Dinucleotide (NAD$^+$)

Humans convert the B vitamin, niacin (also called nicotinic acid), to nicotinamide. Both contain a pyridine ring.

$$\text{Nicotinic acid} + \text{PRPP} + \text{ATP} + \text{glutamine} \longrightarrow \text{NAD}^+ + 2\,\text{PP}_i + \text{glutamate}$$

Except for an additional phosphate that is bound to ribose, the NADP$^+$ coenzyme is identical to the NAD$^+$ coenzyme. These nicotinamide-containing coenzymes are used in dozens of oxidation-reduction reactions:

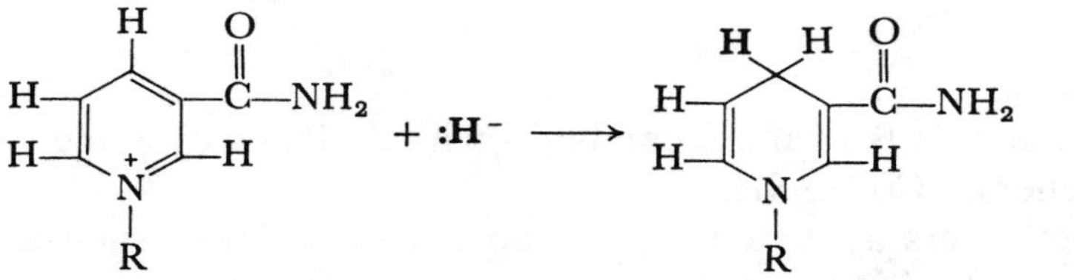

Nicotinamide group of NAD$^+$ Hydride ion   Nicotinamide group of NADH

Humans convert a fraction of their dietary tryptophan to nicotinamide. Thus, the combined niacin-tryptophan intake determines whether enough nicotinamide can be supplied through the diet.

Niacin-deficiency disease, or pellagra, develops in people whose niacin-tryptophan intake is low compared to their caloric intake. Pellagra is classically a disease of people who subsist mainly on corn, which is low in both niacin and tryptophan. Signs of pellagra include dermatitis, diarrhea, and dementia (the three Ds), and loss of tongue papillae.

Major food sources of niacin are meats and nuts.

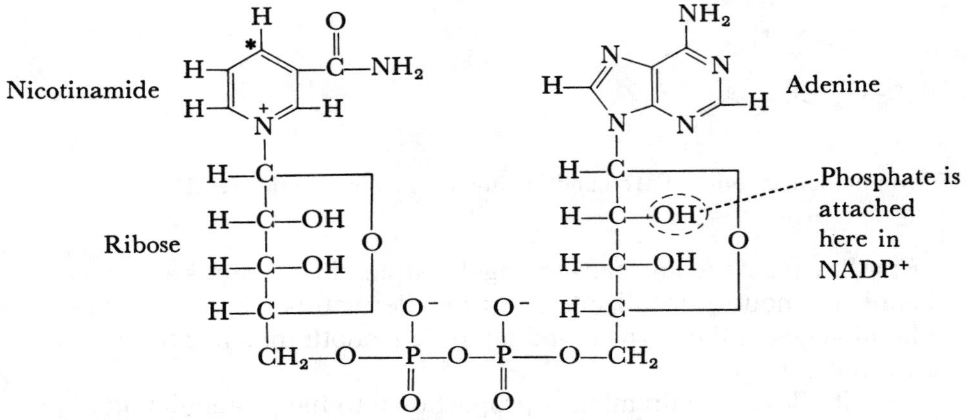

Nicotinamide-adenine dinucleotide (NAD$^+$)

Coenzyme A (CoA)

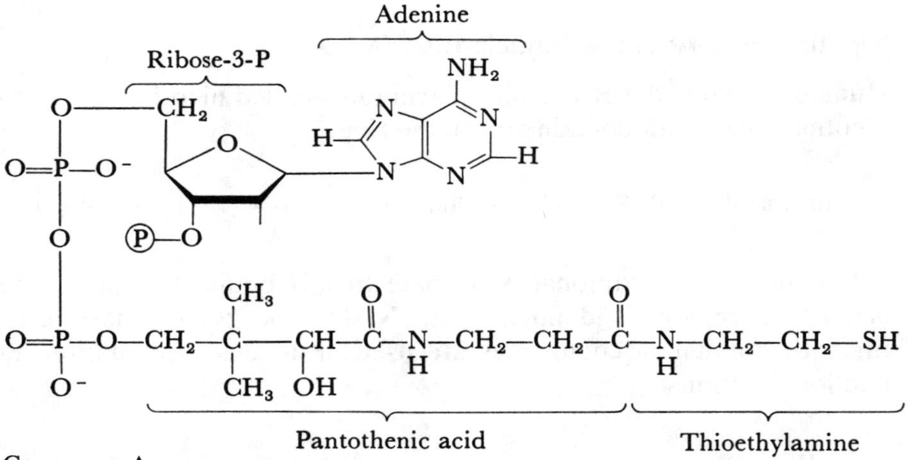

Pantothenic acid, a B vitamin, combines with ATP and cysteine in the liver to generate CoA—SH.

CoA—SH transfers activated acyl groups, R—(C=O)—, such as the acetyl group, by binding them as a thioester. This coenzyme is required by a multitude of different reactions.

Pantothenic acid is found in many foods, and its deficiency is rare.

Pyridoxal Phosphate

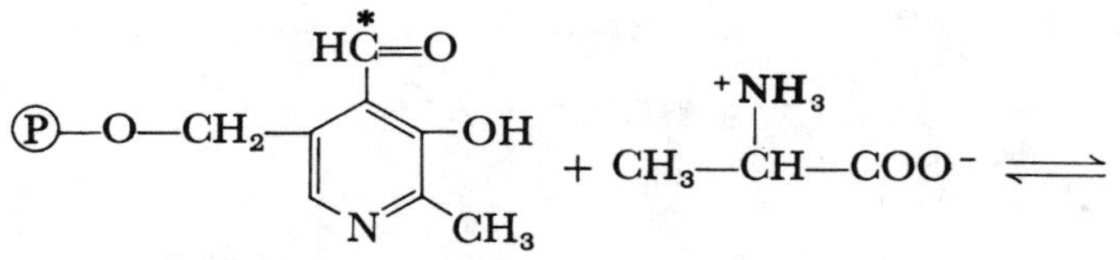

Pyridoxal phosphate   Alanine

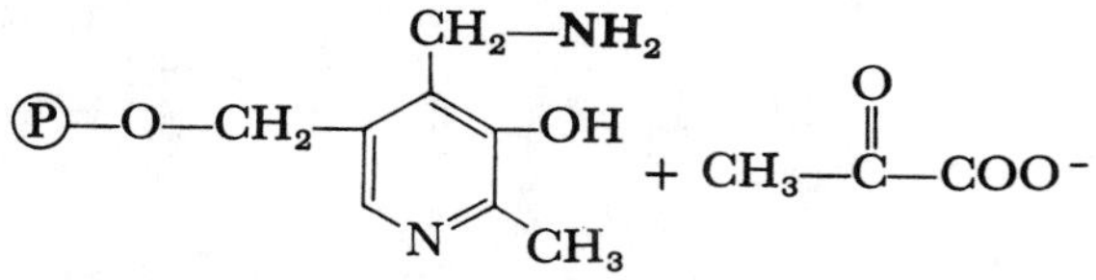

Pyridoxamine phosphate   Pyruvate

Vitamin B_6 includes pyridoxal phosphate, pyridoxamine phosphate, and pyridoxine, the latter two of which humans convert to pyridoxal phosphate. Like nicotinamide, vitamin B_6 contains a pyridine ring.

Pyridoxal phosphate is required to transfer amino groups in transamination reactions and to decarboxylate amino acids. It is the "claw" of amino acid metabolism, in that it brings amino acids into contact with the enzymes that metabolize them.

As shown above, pyridoxal phosphate removes the α-amino group of alanine to produce pyruvate, the corresponding α-keto acid. The pyridoxamine phosphate generated can then donate its amine group to another α-keto acid to transform it into an amino acid.

Vitamin B_6 deficiency is found in many pregnant women and also in alcoholics as well as after chronic administration of B_6 antagonists such as isoniazid and penicillamine. Its features include hypochromic anemia, peripheral neuropathy, irritability, convulsions, and glossitis. Vitamin B_6 deficiency can lead to niacin deficiency, because B_6 is required to convert tryptophan to niacin.

Vitamin B_6 is required in proportion to the protein intake. It may be obtained from a variety of foods.

Biotin

Humans acquire biotin, a B vitamin, both from the diet and from intestinal bacteria. Because of this dual supply, its deficiency is rare. Excessive ingestion of raw egg whites causes biotin deficiency, because avidin, a protein present in raw egg whites, binds to biotin and prevents its absorption. Cooking the eggs denatures avidin.

Biotin is bound to the ε-amino group of the lysine of a carboxylase enzyme. The essential role of biotin is to carry activated carbon dioxide: N-carboxybiotin donates its COO^- group to a substrate to regenerate biotin. Because the carboxyl group in N-carboxybiotin is activated, the transfer of this group occurs without the need for an additional energy source such as ATP.

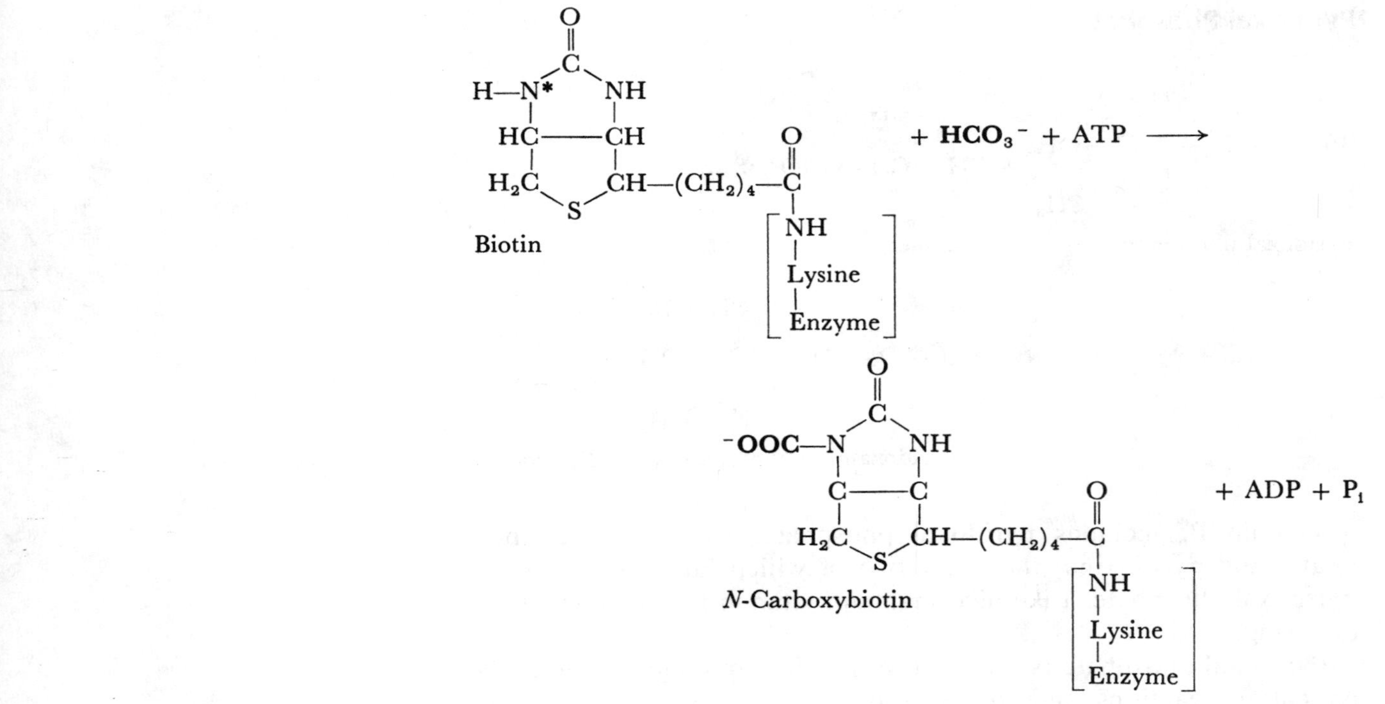

$$\text{Biotin} + HCO_3^- + ATP \longrightarrow$$

$$\textit{N}\text{-Carboxybiotin} + ADP + P_i$$

Tetrahydrofolic Acid (THFA)

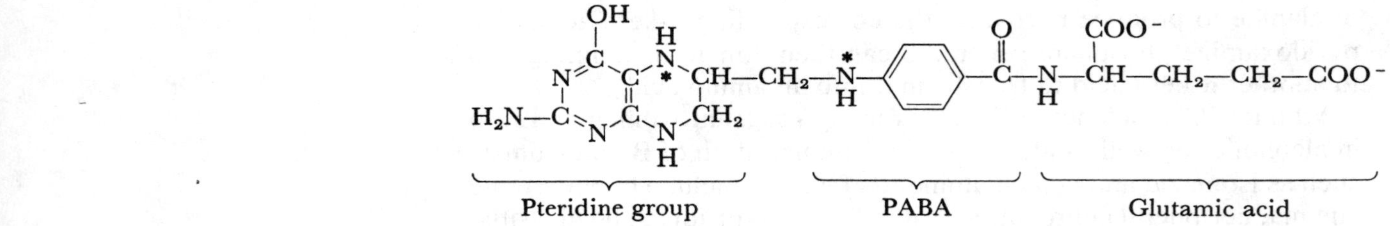

Pteridine group   PABA   Glutamic acid

Tetrahydrofolic acid (THFA)

Folic acid, or pteroylglutamic acid, is a B vitamin that contains a pteridine ring, *p*-aminobenzoic acid (PABA), and glutamate. The principal folate in foods is called pteroylpolyglutamic acid because it contains a chain of glutamate residues that must be deconjugated (cleaved) before absorption. Alcoholics with cirrhosis cannot deconjugate these polyglutamates, and therefore they often develop folate deficiency. After absorption, humans reduce pteroylglutamic acid to THFA, whose structure is shown above. Certain bacteria must synthesize their own folic acid from PABA because they cannot transport pre-formed folate across their cell membranes. Sulfonamide antibiotics act by inhibiting conversion of PABA to dihydrofolic acid (DHFA) within the bacteria. A related antibiotic called trimethoprim works by inhibiting bacterial DHFA reductase, thereby preventing bacterial synthesis of THFA.

The coenzymatic role of both THFA and vitamin B_{12} (cyanocobalamin) is to carry activated one-carbon groups such as methyl (CH_3), methylene (CH$_2$), formyl (CHO), and formimino (CH=NH) groups. This one-carbon pool *does not* include carboxyl groups, which are removed by TPP or pyridoxal phosphate and are added by biotin. Not only does THFA transfer one-carbon groups, it also oxidizes or reduces them. Such one-carbon transfers occur in various pathways, such as in de novo purine synthesis, the conversion of serine to glycine, and the methylation of deoxyuridylic (dUMP) to deoxythymidylic acid (dTMP), an essential step in DNA synthesis:

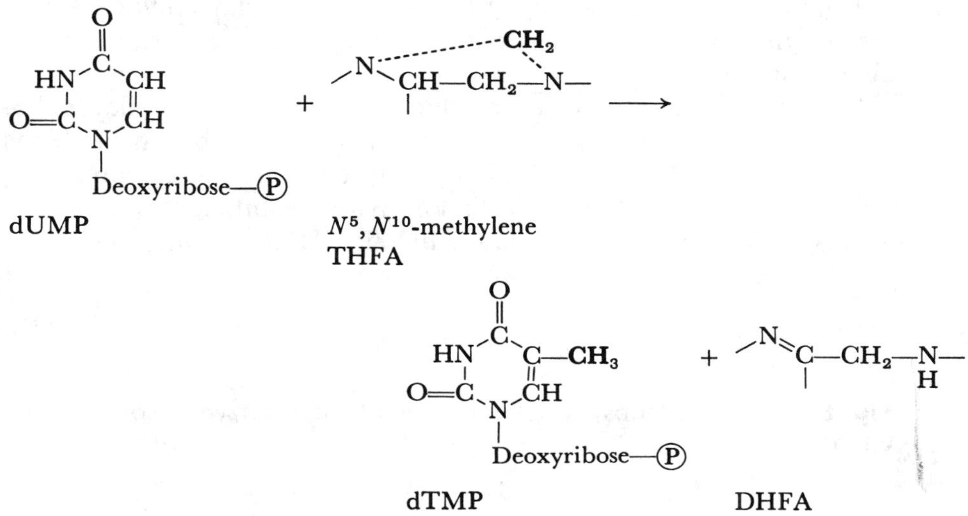

$$\text{dUMP} + N^5,N^{10}\text{-methylene THFA} \longrightarrow$$

dUMP

N^5, N^{10}-methylene
THFA

$$\text{dTMP} + \text{DHFA}$$

dTMP

DHFA

Folate deficiency causes macrocytic, megaloblastic anemia, because it slows both de novo purine synthesis and the conversion of dUMP to dTMP, thereby retarding DNA synthesis.

Foods rich in folic acid include green leafy vegetables, certain fresh fruits, and liver.

Cyanocobalamin

The intestinal absorption of cyanocobalamin, or vitamin B$_{12}$, depends on the gastric secretion of a glycoprotein, termed **intrinsic factor,** which combines with B$_{12}$ and facilitates its absorption in the distal ileum. Without intrinsic factor, very little dietary B$_{12}$ can be absorbed and the vitamin B$_{12}$ excreted into the bile cannot be reabsorbed. The liver stores relatively large amounts of this vitamin. Even when vitamin B$_{12}$ malabsorption occurs, symptoms of deficiency may not develop until the lapse of months or years.

Only microorganisms can synthesize vitamin B$_{12}$. Except for legume nodules, which harbor the B$_{12}$-producing bacteria, plants lack B$_{12}$. Animals obtain the vitamin from microorganisms and from eating other animals. A strict vegeterian diet, which excludes milk and eggs, has virtually no vitamin B$_{12}$. Strict vegetarians in India, however, may continue for decades without developing vitamin B$_{12}$ deficiency, because, with a normal stomach and terminal ileum, they reabsorb most of the B$_{12}$ they excrete into the bile.

After cyanocobalamin is absorbed, the cyanide is removed, and it is converted to the two active cobamide coenzymes: methylcobalamin (methyl-B$_{12}$) and deoxyadenosylcobalamin (DA-B$_{12}$), which is shown in Figure 6-1. The carbon-metal bond with the central cobalt atom of DA-B$_{12}$ is unique. This weak bond can be readily cleaved, generating a 5-deoxyadenosyl radical ($-CH_2$·), which in turn attracts a hydrogen atom from the substrate, creating a substrate radical. This radical then undergoes internal rearrangement, creating a product radical. Finally the B$_{12}$ coenzyme gives the hydrogen atom back to the product radical, yielding the product. Methylmalonyl-CoA mutase requires DA-B$_{12}$ to change the position of its (C$=$O)CoA to create succinyl-CoA.

Vitamin B$_{12}$ deficiency can produce megaloblastic anemia, leukopenia, and thrombocytopenia, which closely resemble the findings in folate deficiency. This similarity may occur because vitamin B$_{12}$ helps convert methyl-THFA to THFA. Without sufficient B$_{12}$, methyl-THFA builds up, and too little methylene-THFA remains to convert dUMP to dTMP.

Fig. 6-1 Deoxyadenosylcobalamin, one of the coenzymes containing vitamine B$_{12}$.

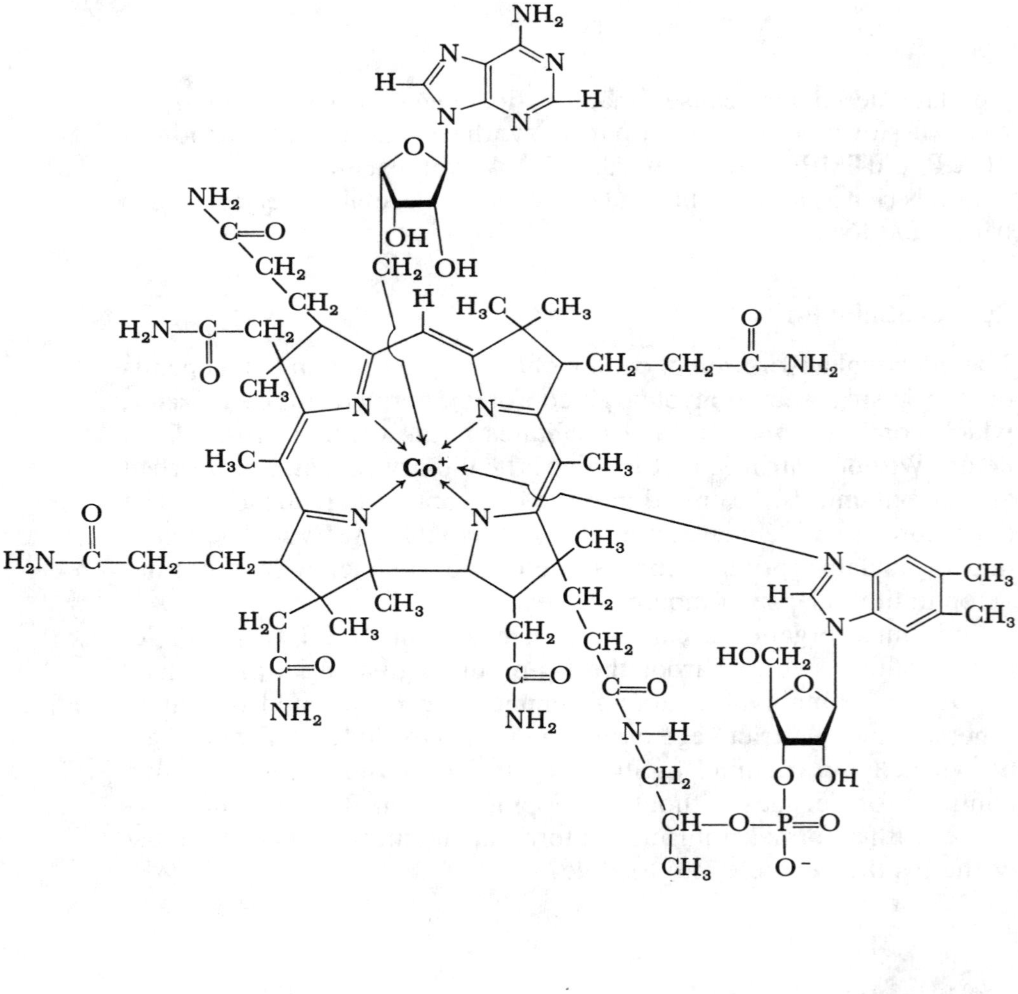

Unlike folate deficiency, vitamin B_{12} deficiency impairs myelin formation by an unknown mechanism, and it can produce peripheral neuropathy and subacute combined degeneration of the spinal cord.

Ascorbic Acid

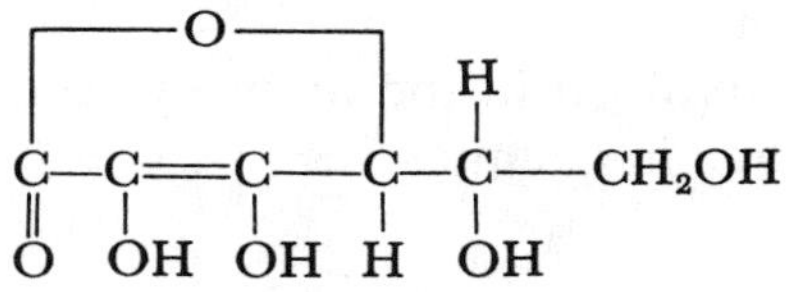

Ascorbic acid

Most species of animals synthesize their own ascorbic acid, or vitamin C, from related hexose sugars. Humans, other primates, and guinea pigs, however, cannot do so, because they lack several enzymes that are necessary to create the essential $—C(OH){=}C(OH)—$ of ascorbate.

Ascorbic acid functions both as an antioxidant and as a coenzyme utilized in oxygenation reactions. Ascorbate acts as a reducing agent, maintaining the iron atom of enzymes in the ferrous rather than ferric state. Ascorbate supports reactions such as the oxygenation of proline and the hydroxylation of proline and lysine in collagen. Ascorbate also improves intestinal iron absorption.

The features of scurvy (severe vitamin C deficiency) include follicular hyperkeratosis, petechiae, subconjunctival hemorrhage, gum changes, subperiosteal hemorrhage, and, in children, a failure to grow.

The major food sources of vitamin C are fresh fruit and vegetables.

Coenzymes Not Containing Vitamins

Not all coenzymes contain vitamins, as exemplified by coenzyme Q (the ubiquinones) and lipoic acid.

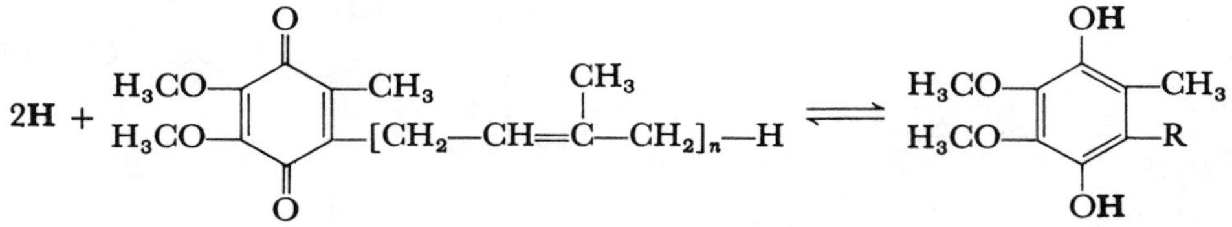

Ubiquinone (oxidized)   Ubiquinol (reduced)

Coenzyme Q (CoQ) transfers H atoms and electrons in the mitochondrial oxidative-phosphorylation system. Humans synthesize their own ubiquinones.

Reduced lipoic acid is a short-chain fatty acid with two sulfhydryl groups. After oxidation these groups form a disulfide linkage, as shown previously in the discussion of thiamine (TPP) metabolism.

Although lipoic acid is itself insoluble in water, it forms water-soluble salts.

Two major reactions that utilize lipoic acid are the pyruvate and α-ketoglutarate dehydrogenase reactions, each of which uses TPP, lipoic acid, CoA, FAD^+, and NAD^+.

No lipoic acid deficiency state has been observed in humans; hence, lipoic acid is a coenzyme but not a vitamin.

Tetrahydrobiopterin is a coenzyme involved in the hydroxylation of phenylalanine to tyrosine, as described in Chapter 12.

Vitamins Not Acting as Coenzymes— The Fat-Soluble Vitamins
Vitamin A

Humans ingest two types of vitamin A: provitamin A from plants and pre-formed vitamin A from animal sources.

β-Carotene, the principal dietary provitamin A, abounds in yellow and orange vegetables. As shown below, each molecule of β-carotene is cleaved in the intestinal mucosa to two molecules of retinol, or vitamin A_1.

The main sources of pre-formed vitamin A, or retinol, are liver, whole milk, fish oils, and eggs. Muscle meats and plants do not contain pre-formed vitamin A.

Vitamin A_1 is oxidized to retinol, or vitamin-A aldehyde, which combines with opsin, a protein, to form rhodopsin, the light-sensing pigment in the retina. Thus, the earliest symptom of vitamin A deficiency is night blindness.

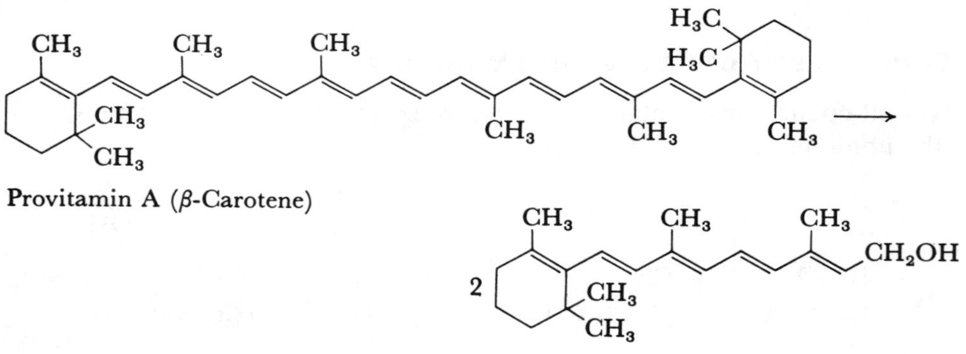

Provitamin A (β-Carotene)

Vitamin A_1 (Retinol)

In addition, vitamin A is required to form and maintain epithelial surfaces through a mechanism that is still unknown. Vitamin A deficiency causes follicular hyperkeratosis (i.e., the development of keratin plugs in hair follicles, as seen in scurvy) and a xerophthalmia (i.e., corneal dryness) that can progress to corneal ulcers and resultant blindness.

Acute vitamin A intoxication has occurred in Arctic explorers who ate polar bear livers. Chronic hypervitaminosis A usually occurs after enormously excessive vitamin A ingestion by food faddists or in the treatment of acne. Its features may include arthralgias, fatigue, skin changes, and headaches due to benign intracranial hypertension (pseudotumor cerebri).

Though harmless, excessive β-carotene ingestion makes the skin yellow or orange. In distinction to the observation in cases of jaundice, the sclera remains white.

Vitamin D

Vitamin D is called the "solar vitamin," because its synthesis involves the ultraviolet irradiation of sterols, such as 7-dehydrocholesterol in human skin (shown below), or of ergosterol from milk.

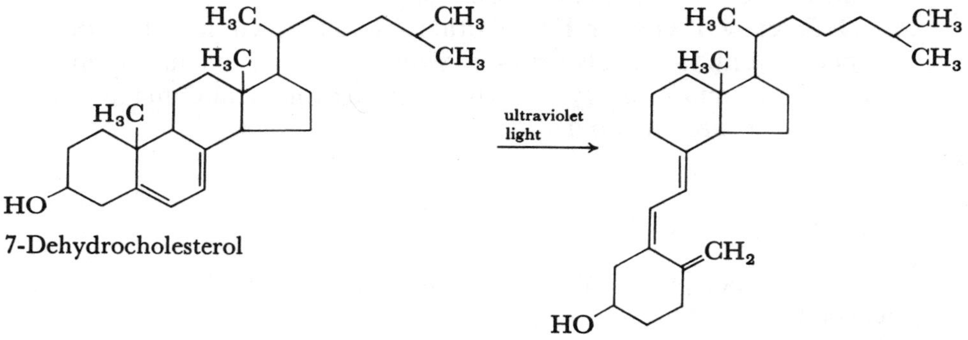

7-Dehydrocholesterol

Cholecalciferol
(Vitamin D_3)

The only foods that naturally contain vitamin D are fresh oils, such as fish liver oils. Milk becomes a source of vitamin D only after irradiating it or adding vitamin D. Humans exposed to bright sunlight year-round do not require dietary vitamin D.

Cholecalciferol, or vitamin D_3, must be hydroxylated twice to become metabolically active. It is hydroxylated first in the liver to produce 25-hydroxycholecalciferol (25-HCC), and next in the kidney to generate calcitriol (1,25-dihydroxycholecalciferol, or 1,25-DHCC), the most active form of vitamin D. Calcitriol acts as a hormone. Its principal role is to facilitate intestinal calcium absorption. In addition, it aids parathyroid hormone in mobilizing bone calcium.

Vitamin D deficiency in children, known as rickets, is most common in areas lacking sunshine. Rickets deforms the growing bones of children far more than does vitamin D deficiency in adults, which is termed osteomalacia.

Hypervitaminosis D occurs only after chronic, greatly excessive vitamin D administration. It produces hypercalcemia with resultant widespread calcification and kidney stone formation.

Vitamin E

Vitamin E consists of a group of tocopherols, the α-tocopherols being the most active.

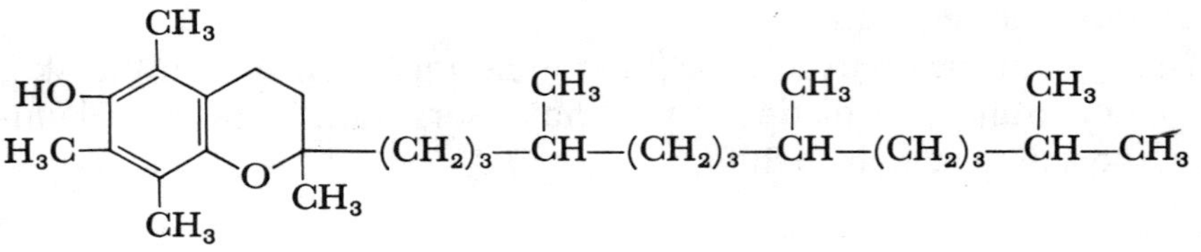

α-Tocopherol

The main dietary source of vitamin E is vegetable oil. It also occurs in grains and leafy vegetables.

The only established role of vitamin E in humans is to protect the polyunsaturated fats and vitamin A from oxidation. As an antioxidant, it also protects the erythrocytes.

Deficiency of vitamin E (and other fat-soluble vitamins) is common in people with fat malabsorption syndromes and can cause hemolysis (rapid destruction of erythrocytes) and neurological damage such as impaired sensation and gait.

Vitamin K

Vitamin K is a group of naphthoquinones with long, branched hydrocarbon side chains.

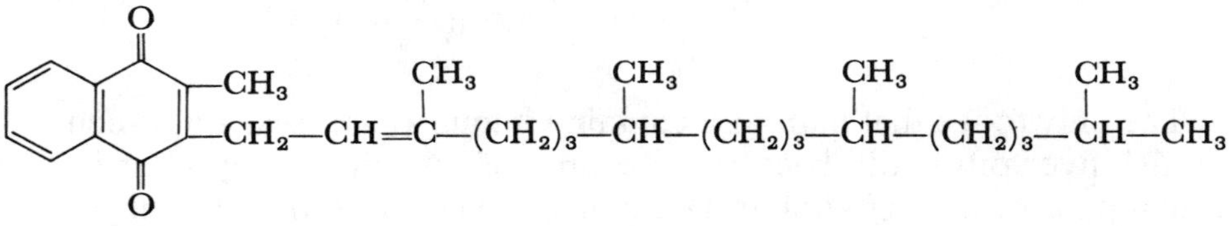

Vitamin K

The main dietary source of vitamin K is from green leafy vegetables. In addition, intestinal bacteria synthesize this vitamin.

Vitamin K acts in the liver to promote the carboxylation of glutamate in prothrombin, thereby activating this coagulation factor. Deficiency will reduce the plasma concentrations of these clotting factors and predispose to hemorrhage. Vitamin K deficiency can occur during the first few days after birth, because newborns lack the intestinal bacteria that produce vitamin K and because they have no store of vitamin K (it does not cross the placenta). Hence, all newborns are given a vitamin K injection to prevent hemorrhagic disease. Vitamin K deficiency may also occur following antibiotic therapy that sterilizes the gut.

Dicumarol drugs are important oral anticoagulants that act by antagonizing the action of vitamin K in the liver.

Problems

Problem 1

Which of the following coenzymes is not derived from vitamins?

    A. CoA—SH.
    B. Pyridoxal phosphate.
    C. TPP.
    D. Methylcobalamin.
    E. Lipoamide.

Problem 2

Tetrahydrofolic acid:

    A. Is an important antioxidant.
    B. Often acts in reactions requiring thiamine pyrophosphate.
    C. Transfers activated CO_2.
    D. Is important in vitamin B_{12} metabolism.
    E. Is essential for the hydroxylation of phenylalanine to tyrosine.

Problem 3

Coenzymes:

    A. Alter the equilibrium of reactions.
    B. Are consumed by reactions.
    C. Usually consist of polypeptides.
    D. Often transfer activated groups.
    E. Include Mg^{+2}, Zn^{+2}, and Fe^{+2}.

Problems 4–6

Match the substances below to the reactions in Problems 4–6. Each problem may have more than one correct response.

    A. Pantothenic acid
    B. Biotin
    C. Lipoamide
    D. Tetrahydrofolic acid
    E. Thiamine
    F. Ascorbic acid

4. Decarboxylation of α-ketoglutarate to form succinyl-CoA.
5. Transfer of a formyl group to methionine.
6. Carboxylation of pyruvate to oxaloacetate.

Answers

1. E.
2. D. Biotin carries activated CO_2.
3. D. Coenzymes are not polypeptides. Like enzymes, they do not alter the equilibrium of a reaction nor are they consumed by the reaction.
4. A, C, E.
5. D. Tetrahydrofolic acid transfers one-carbon formyl groups.
6. B.

References

Devlin, T. M. *Textbook of Biochemistry with Clinical Correlations* (3rd ed.). New York: Wiley-Liss, 1992. Pp. 1115–1145.

Mathews, C. K., and van Holde, K. E. *Biochemistry*. Redwood City, Calif.: Benjamin/Cummings, 1990. Pp. 365–372, 474–478, 587, 635–636, 691–702.

Murray, R. K., Granner, D. K. Mayes, P. A., and Rodwell, V. W. *Harper's Biochemistry* (22nd ed.). Norwalk, Conn.: Appleton & Lange, 1990. Pp. 58–60, 547–570.

Stryer, L. *Biochemistry* (3rd ed.). New York: Freeman, 1988. Pp. 259, 268, 322–324, 379–382, 403, 434, 440–441, 496–499, 506–509, 569–571, 580–583, 617–618.

Structure and Properties of Carbohydrates

Carbohydrates are defined as polyhydroxylated compounds with at least three carbon atoms that may or may not possess a carbonyl group. The formulas of many carbohydrates show one water molecule for every carbon; hence, the name "carbohydrate."

The aldehyde sugars are called **aldoses,** while those with a ketone group are **ketoses.** In designating their relative position, the carbon atoms are numbered so as to assign the lowest possible number to the carbonyl carbon.

Sugar alcohols contain a hydroxyl group in place of a carbonyl group. Mannitol, the sugar alcohol derived from mannose, is frequently used medically as an osmotic diuretic to reduce cerebral edema. Sorbitol, the sugar alcohol derived from glucose, often accumulates in the lenses of diabetics and produces cataracts.

Sugar acids, such as glucuronic and ascorbic acids, contain a carboxyl group.

Other groups that can substitute for a hydroxyl group include phosphate, sulfate, amino, and *N*-acetyl groups. The deoxy sugars of DNA lack a hydroxyl group at carbon 2.

Trioses

The three-carbon sugars, the **trioses,** are the smallest possible carbohydrates. Several trioses are shown on page 60 according to the Fischer method of representation. If the OH group on the penultimate (next to last) carbon atom points to the right (**dextro**) on the Fischer structure, the sugar is the D isomer. If the OH group points to the left (**levo**), it is the L isomer of the sugar. These designations have nothing to do with optical isomerization—i.e., the ability to rotate the plane of polarized light—which is denoted by *d* or *l*. The D and L isomers of the same sugar are known as enantiomers: mirror images of one another.

Glyceraldehyde is an aldotriose, dihydroxyacetone (DHA) is a ketotriose, and glycerol is a sugar alcohol, as shown on page 60. Lactic and pyruvic acids are derived from glyceraldehyde and DHA, respectively; neither fits the definition of a carbohydrate, because each has only one hydroxyl group.

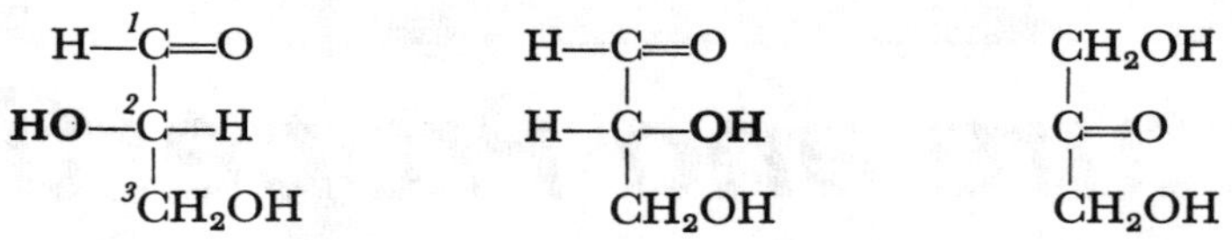

$$H-\overset{1}{C}=O$$
$$HO-\overset{2}{C}-H$$
$$\overset{3}{C}H_2OH$$
L-Glyceraldehyde

$$H-C=O$$
$$H-C-OH$$
$$CH_2OH$$
D-Glyceraldehyde

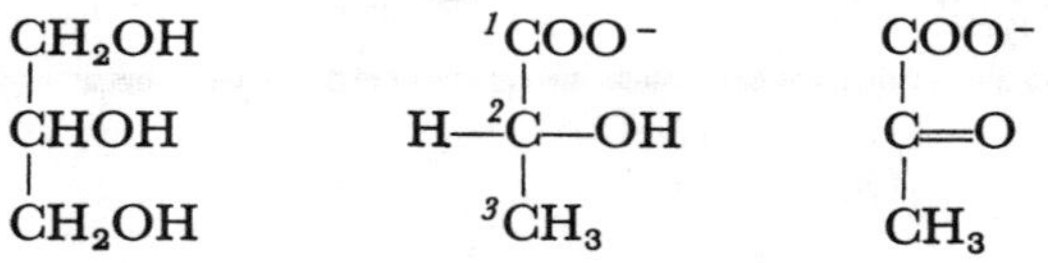

$$CH_2OH$$
$$C=O$$
$$CH_2OH$$
DHA

$$CH_2OH$$
$$CHOH$$
$$CH_2OH$$
Glycerol

$$\overset{1}{C}OO^-$$
$$H-\overset{2}{C}-OH$$
$$\overset{3}{C}H_3$$
D-Lactic acid

$$COO^-$$
$$C=O$$
$$CH_3$$
Pyruvic acid

Tetroses and Pentoses

Four-carbon sugars, or **tetroses,** play a minor role in humans compared to the five-carbon (**pentose**) and six-carbon (**hexose**) sugars.

The major human pentoses are ribose and 2-deoxyribose. In solution, each exists primarily as a five-membered ring that contains oxygen and four carbons. Since this ring resembles that of furan, these sugars are said to be in furanose form.

The Haworth projections shown below place the furanose ring perpendicular to the plane of the page. If an OH group projects to the right on the Fischer structure, it will project downward on the Haworth projection.

The carbonyl carbon atom of pentoses and hexoses is termed the **anomeric** carbon atom. Ring formation links the anomeric carbon to the penultimate carbon and transforms it into an asymmetric carbon atom with two possible configurations. In the α-anomer, the OH group that is attached to this carbon atom is to the right on the Fischer structure and down on the Haworth projection. The β-anomer has the OH projecting to the left on the Fischer structure and upward on the Haworth projection. An asterisk denotes the anomeric carbon on the structures below.

The spontaneous interconversion of α and β anomers in solution is termed **mutarotation.**

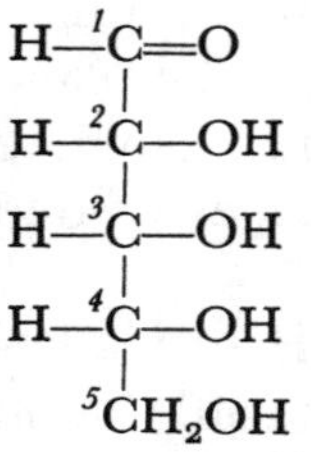

$$H-\overset{1}{C}=O$$
$$H-\overset{2}{C}-OH$$
$$H-\overset{3}{C}-OH$$
$$H-\overset{4}{C}-OH$$
$$\overset{5}{C}H_2OH$$
D-Ribose, open chain
(Fischer structure)

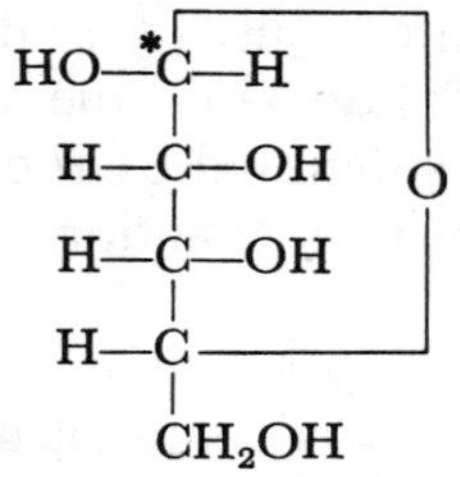

$$HO-\overset{*}{C}-H$$
$$H-C-OH$$
$$H-C-OH$$
$$H-C$$
$$CH_2OH$$
β-D-Ribofuranose
(Fischer structure)

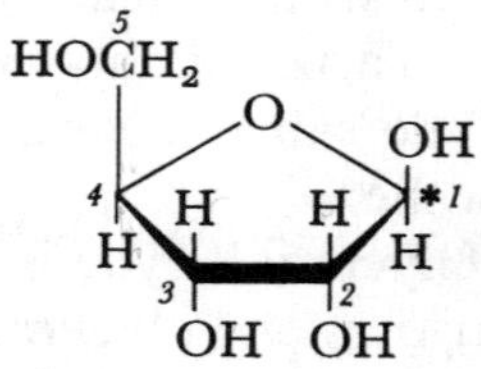

β-D-Ribofuranose
(Haworth projection)

Hexoses

The two major **aldohexoses** in humans, glucose and galactose, differ only in the configuration of their OH group at carbon 4; they are **epimers.** In solution, less than 1% of either remains as a straight chain. Instead, each forms a six-membered ring that contains five carbon atoms and one oxygen. Because this ring resembles that of pyran, these sugars are termed pyranoses.

The major **ketohexose** in humans is fructose. Fructose exists mainly in the furanose form with a small amount in the pyranose form, and it has carbon 2 as its anomeric carbon atom.

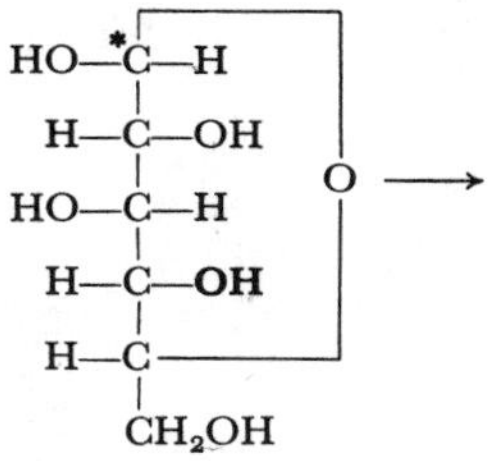

β-D-Glucopyranose 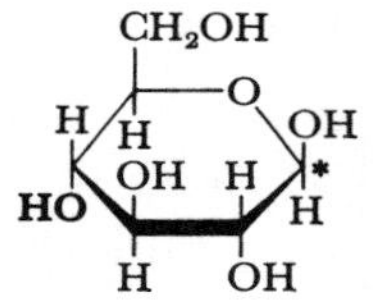 β-D-Glucopyranose (Haworth projection) 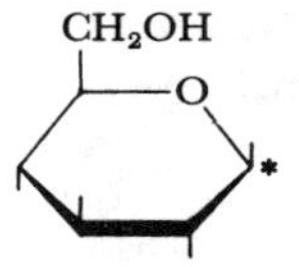 β-D-Glucopyranose (simplified Haworth projection)

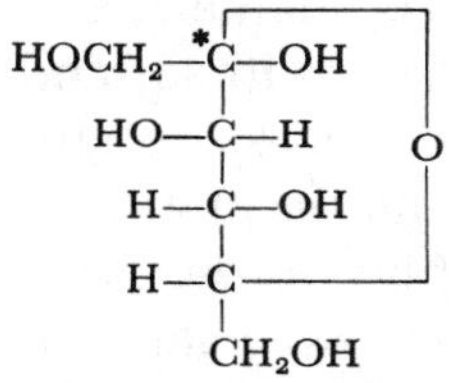

α-D-Fructofuranose 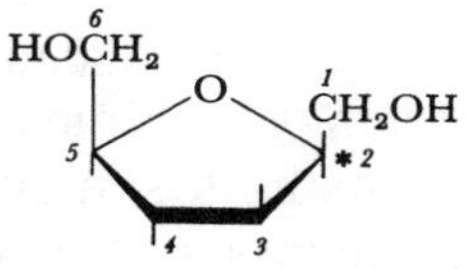 α-D-Fructofuranose 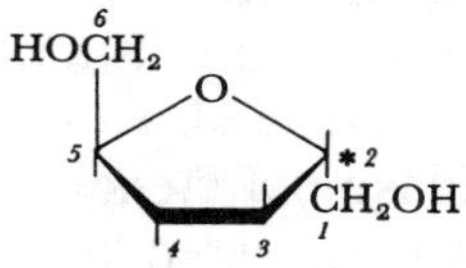 β-D-Fructofuranose

The actual conformation of the pyranose and furanose rings is not planar, as the Haworth projection makes it appear. Instead, the pyranose ring exists mainly in "chair" form, shown below with (left) and without (right) the substituents bound to the carbon atoms.

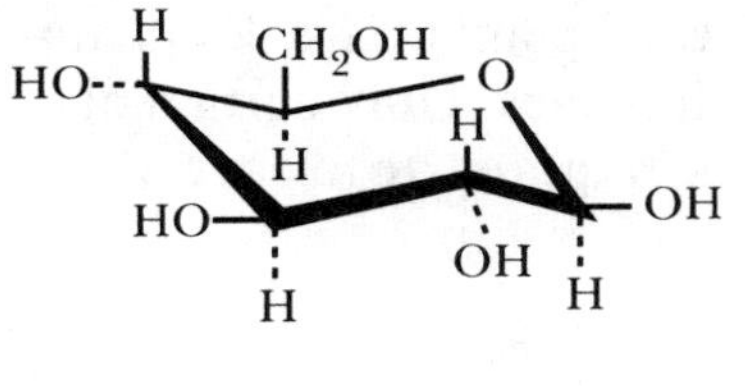

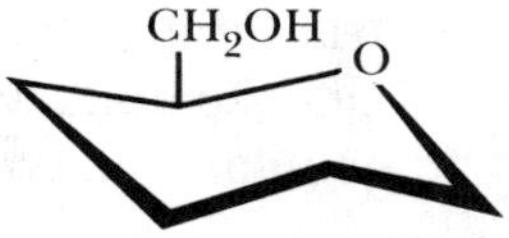

Disaccharides

Disaccharides consist of two monosaccharides joined by a glycosidic bond. Each glycosidic bond is classified as α or β and is numbered according to the positions of the carbon atoms it links.

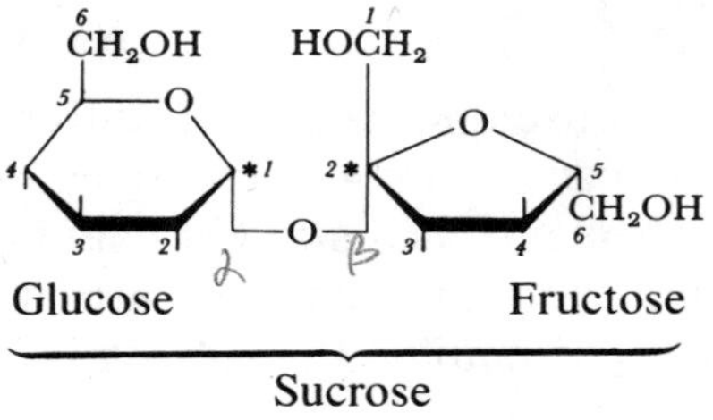

Sucrose

In sucrose, the α-anomeric carbon 1 of glucose joins the β-anomeric carbon 2 of fructose.

Maltose, a dimer of glucose linked by an α-1,4 glycosidic bond, is produced during gastrointestinal starch digestion.

Lactose, found naturally only in milk products, consists of β-galactose with a β-1,4 linkage to glucose.

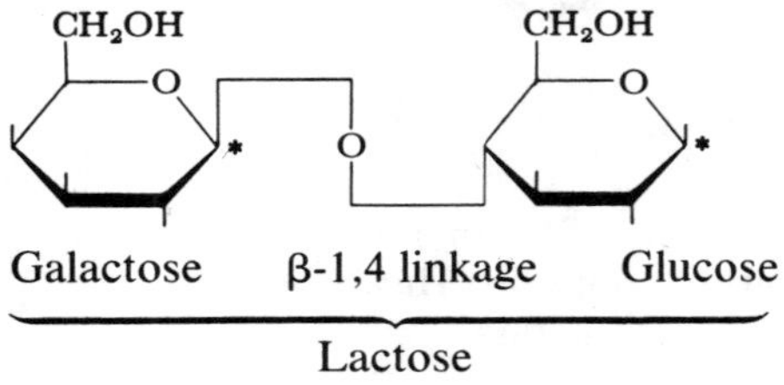

Lactose

The intestinal mucosa contains the enzymes sucrase, maltase, and lactase, which cleave the glycosidic bonds of sucrose, maltose, and lactose, respectively. Sucrase, for example, cleaves sucrose to glucose and fructose. Lactase deficiency, the most common intestinal disaccharidase deficiency, causes osmotic diarrhea (due to the presence of the undigested lactose in the intestine) and excessive gas production as a result of lactose fermentation by intestinal bacteria.

Polysaccharides

Polysaccharides are carbohydrate polymers with more than 10 monosaccharide units. **Oligosaccharides** contain 2 to 10 monosaccharide residues.

Cellulose, the structural polysaccharide of plants, is a long, unbranched chain of glucose units with β-1,4 linkages. Only certain bacteria possess the cellulase enzyme that is required to cleave β-1,4

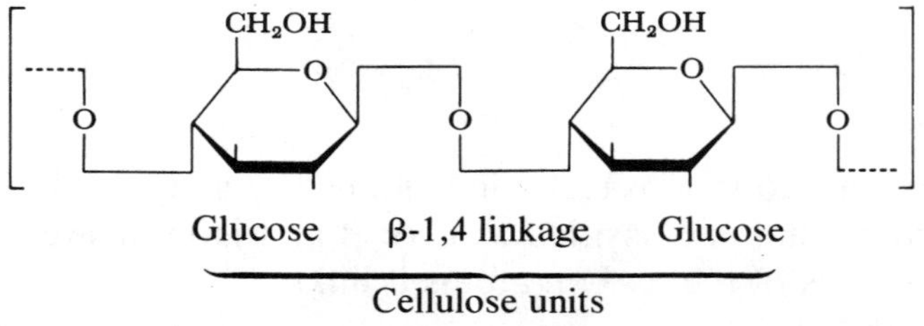

Cellulose units

glycosidic bonds. Cows depend on bacteria in their rumen to digest cellulose.

Starch, the energy-storing polysaccharide of plants, is a mixture of amylose and amylopectin. Amylose is a long, unbranched glucose polymer with α-1,4 bonds. Amylopectin, on the other hand, has α-1,6 branch linkages spaced about every 30 glucose residues in its α-1,4 chain (Fig. 7-1).

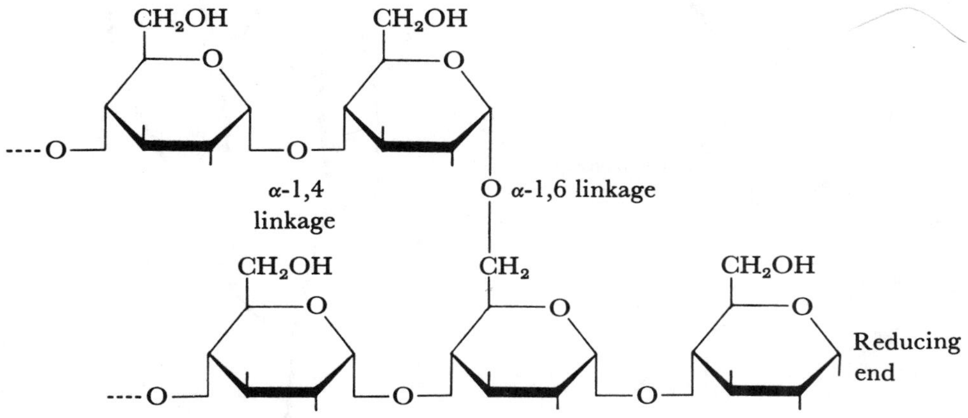

Amylopectin

Glycogen, the glucose-storing polysaccharide of animals, resembles amylopectin in having α-1,6 branches from an α-1,4 chain. Glycogen, however, is more highly branched, with α-1,6 linkages about every 10 glucose residues, as shown in Figure 7-1.

1) α-Amylase, produced by the salivary glands and pancreas, randomly hydrolyzes the α-1,4 linkages of dietary polysaccharides. α-Amylase hydrolyzes amylose to a mixture of maltose and maltotriose (a triose composed of three glucose residues joined by α-1,4 linkages). 2) Maltase then hydrolyzes maltose and maltotriose to glucose. When α-amylase digests dietary amylopectin and glycogen, it cannot attack the α-1,6 linkages. Thus, after cleaving the branch chains down to the α-1,6 bonds, it leaves a glucose polymer skeleton called a **limit dextrin.**

3) α-Dextrinase, the debranching enzyme, hydrolyzes these α-1,6 bonds and allows α-amylase to continue its task on the internal α-1,4 linkages. 4)

Glycogen phosphorylase in human tissues removes glucose from tissue glycogen by phosphorolysis to glucose-1-phosphate. Glucan transferase then transfers the final several glucose residues of the branch to another arm of the polymer, leaving a limit dextrin as shown in Figure 7-2. Amylo-1,6-glucosidase must then debranch this limit dextrin to allow glycogen phosphorylase to degrade the remainder of the glycogen molecule to glucose-1-phosphate.

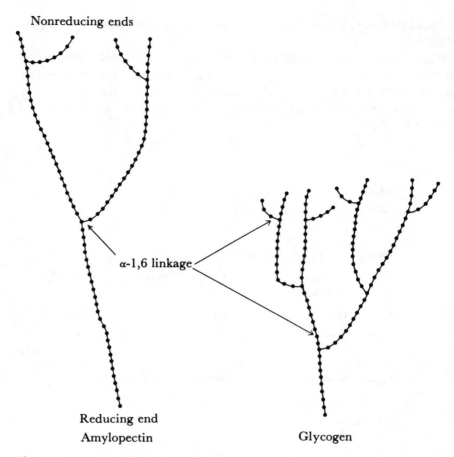

Fig. 7-1 Comparison of amylopectin and glycogen. Amylopectin has branches about every 30 glucose units, whereas glycogen has branches about every 10 glucose residues.

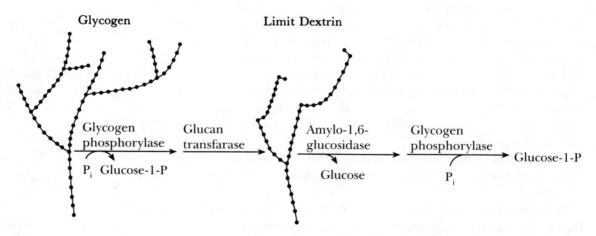

Fig. 7-2 Degradation of glycogen. Glycogen phosphorylase, glucan transferase, and amylo-1,6-glucosidase convert glycogen into a limit dextrin.

Glycoproteins

Also known as proteoglycans, glycoproteins were formerly called mucopolysaccharides. Glycoproteins consist mainly of carbohydrates with a small amount of protein. Glycoproteins abound in the ground substance of connective tissue, where they play a major structural role. Loss of glycoproteins from intervertebral discs predisposes to degenerative disc disease. The carbohydrate components of glycoproteins are called glycosaminoglycans. Built from repeating disaccharide units, the glycosaminoglycans are shaped as long monomers. The sugar components of the disaccharide units consist of an amino sugar plus, in all but keratan sulfate, a sugar acid. These sugar components often contain N-acetyl or sulfate groups.

Hyaluronic acid is an acidic glycosaminoglycan made up of the repeating disaccharide unit of glucuronic acid joined to N-acetylglucosamine. The staphylococci, for example, can readily invade connective tissue, because they secrete hyaluronidase, an enzyme that cleaves hyaluronic acid.

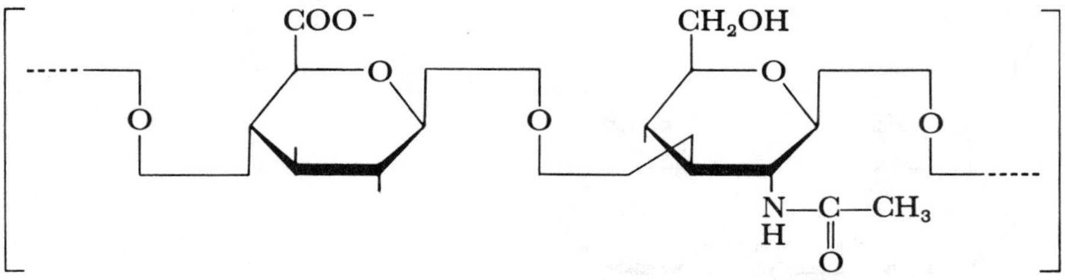

Glucuronic acid N-Acetylglucosamine

Repeating unit of hyaluronic acid

Heparin, a sulfated, acidic glycosaminoglycan, is normally present in mast cells and is commonly administered as an anticoagulant. It consists of an O-sulfated glucuronic or iduronic acid bound to an N-sulfated glucosamine that has a second O-sulfate group.

Chondroitin sulfate A contains the repeating disaccharide unit of glucuronic acid joined to N-acetylgalactosamine-4-sulfate.

The somatomedins are a group of hormones that promote the sulfation of glycoproteins. Somatomedin deficiency impairs growth.

The mucopolysaccharidoses are a class of rare, inherited disorders of glycoprotein degradation. Deficiency of enzymes that break down glycosamino-glycan leads to the accumulation of these components, resulting in bony deformities and abnormal facial appearance.

Problems

Problems 1–4

Match the disaccharide structures in Problems 1–4 to the list below:

    A. Lactose
    B. Glycosaminoglycan
    C. Sucrose
    D. Maltose

1. C

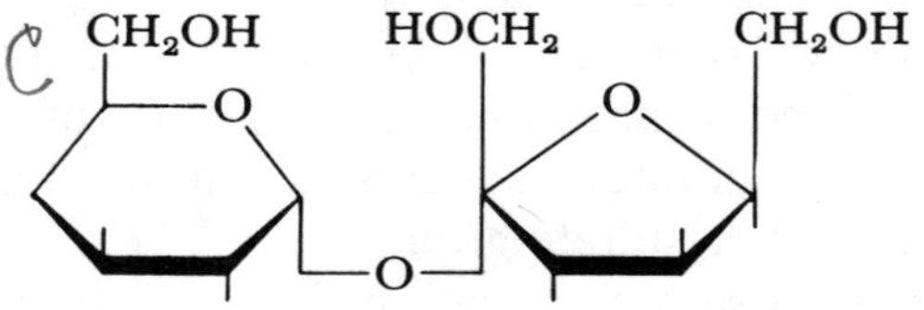

2. B

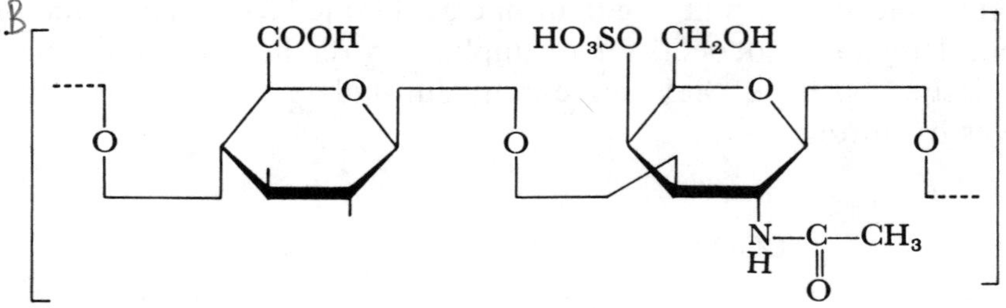

3. D

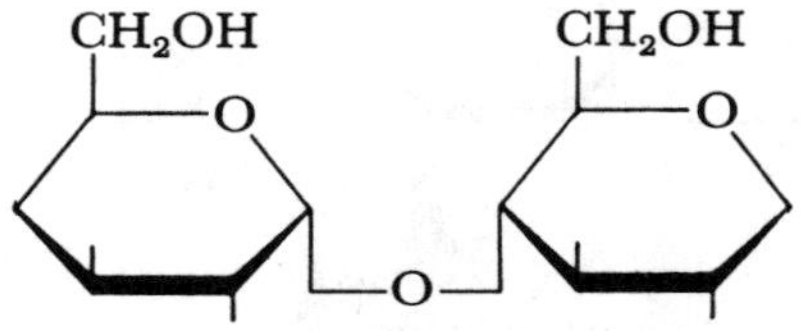

4. B

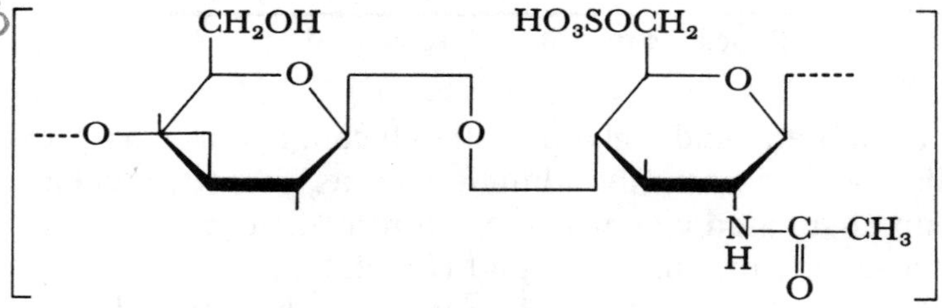

Problem 5

Which of the following is cleaved to maltose and maltotriose following hydrolysis by α-amylase?

    A. Glycogen.
    B. Cellulose.
    C. Amylopectin.
    D. Keratan sulfate.
    √E. Amylose.

Problem 6

A person deficient in glucan transferase:

    A. Can completely degrade glycogen using glycogen phosphorylase and amylo-1,6-glucosidase.

B. Accumulates unbranched glycogen in the liver.
C. Cannot metabolize the glucose-1-phosphate produced by glycogen phosphorylase action on glycogen.
D. Has excess glycogen and limit dextrins in the liver.
E. Is intolerant of lactose.

Answers

1. C. Glucose bound to fructose.
2. B. A sugar acid bound to a sulfated amino sugar.
3. A. Galactose bound to glucose.
4. B. Galactose bound to *N*-acetylglucosamine-6-sulfate.
5. E.
6. D. Deficiency of this debranching enzyme leads to the accumulation of limit dextrins and glycogen. Unbranched glycogen is not produced. The conversion of glucose-1-phosphate to glucose-6-phosphate is not affected. Lactose intolerance is produced by lactase deficiency.

References

Devlin, T. M. *Textbook of Biochemistry with Clinical Correlations* (3rd ed.). New York: Wiley-Liss, 1992. Pp. 337–343, 1151–1154.

Mathews, C. K., and van Holde, K. E. *Biochemistry*. Redwood City, Calif.: Benjamin/Cummings, 1990. Pp. 260–296.

Murray, R. K., Granner, D. K., Mayes, P. A., and Rodwell, V. W. *Harper's Biochemistry* (22nd ed.). Norwalk, Conn.: Appleton & Lange, 1990. Pp. 124–133, 591–608.

Stryer, L. *Biochemistry* (3rd ed.). New York: Freeman, 1988. Pp. 275–277, 298–299, 331–348, 449–453.

8

Carbohydrate Catabolism and Biosynthesis

Dietary carbohydrate consists mainly of the polysaccharides, amylose and amylopectin, and the disaccharides, sucrose and lactose, with small amounts of free glucose and fructose. The principal monosaccharide derived from the intestinal hydrolysis of dietary carbohydrate is glucose. Fructose and galactose appear in lesser amounts, and they must be especially channeled into the mainstream of glucose breakdown, called **glycolysis** (lysis of glucose).

Fructose Metabolism

Most human tissues cannot utilize fructose. The principal organs that metabolize fructose are the liver, kidneys, intestine, and adipose tissue.

The main enzyme phosphorylating glucose in the liver, glucokinase, cannot phosphorylate fructose to fructose-6-phosphate (fructose-6-P).

Although hexokinase can phosphorylate fructose to fructose-6-P, the enzyme has a much higher affinity for glucose than fructose. Thus, the relative abundance of glucose in the liver competitively inhibits conversion of fructose to fructose-6-P. Adipose tissue, however, contains more fructose than glucose. In consequence, hexokinase in adipose tissue produces fructose-6-P, an intermediate in glycolysis.

The liver has another enzyme, fructokinase, which phosphorylates fructose to fructose-1-P. Fructose-1-P aldolase, in turn, splits this ketohexose-1-P into dihydroxyacetone phosphate (DHAP), a ketotriose-1-P, and glyceraldehyde, an aldotriose. DHAP and glyceraldehyde can then proceed through glycolysis to yield energy, or they can be converted to glucose, which can then be utilized by all the tissues of the body.

Fructose
    ↓ Fructokinase

Fructose-1-P
    ↓ Fructose-1-P aldolase

DHAP + Glyceraldehyde

Hereditary fructokinase deficiency causes no symptoms, but affected individuals will spill fructose into their urine. This fructosuria might be misinterpreted as glucosuria, because both yield a positive reducing-sugar test.

Fructose-1-P aldolase deficiency, called **hereditary fructose intolerance,** causes fructose-1-P to accumulate inside liver cells after sucrose or fructose ingestion. This accumulation results in hypoglycemia (low serum glucose levels) and vomiting. The treatment of hereditary fructose intolerance is avoidance of dietary fructose and sucrose.

Galactose Metabolism

Like fructose, only a few organs can metabolize galactose; the most important of these are the liver and erythrocytes.

Galactose metabolism requires the formation of uridine diphosphate-glucose (UDP-glucose), which is created from glucose-1-P and UTP:

$$\text{Glucose-1-P} + \text{UTP} \rightleftharpoons \text{UDP-glucose} + \text{PP}_i$$

Galactokinase phosphorylates galactose to galactose-1-P. Next, a uridyl transferase removes glucose-1-P from UDP-glucose and creates UDP-galactose.

Humans have two uridyl transferases: galactose-1-P and hexose-1-P uridyl transferase. Finally, UDP-glucose epimerase changes the OH configuration at carbon 4′ of UDP-galactose to yield UDP-glucose, an intermediate in glycogen synthesis:

$$\text{Galactose} + \text{ATP} \underset{\text{galactokinase}}{\rightleftharpoons} \text{galactose-1-P} + \text{ADP}$$

$$\text{Galactose-1-P} + \text{UDP-glucose} \underset{\text{uridyl transferase}}{\rightleftharpoons} \text{UDP-galactose} + \text{glucose-1-P}$$

$$\text{UDP-galactose} \underset{\text{epimerase}}{\rightleftharpoons} \text{UDP-glucose}$$

Galactosemia, an increased level of serum galactose, has either of two causes: hereditary galactokinase deficiency or hereditary galactose-1-P uridyl-transferase deficiency. In both disorders, galactose is reduced to its sugar alcohol, galactitol, which may be deposited inside the lens of the eye. Galactitol increases the osmotic pressure inside the lens and can cause cataracts. In only the uridyl-transferase deficiency is galactose-1-P trapped within the liver cells and erythrocytes, leading to hepatomegaly, impaired liver function, and mental retardation if the dietary galactose intake continues. (The uridyl-transferase deficiency resembles hereditary fructose intolerance. In both, a hexose-1-P cannot be metabolized nor can it leave the liver cells because of its charge; hence, it accumulates and causes hepatomegaly and liver dysfunction.) The treatment for both types of galactosemia is to eliminate milk products, the source of lactose, which contains galactose.

Glycolysis

Glycolysis is the main pathway for carbohydrate catabolism in virtually every human tissue (Fig. 8-1). In studying glycolysis, particular focus should be placed on the two ATP-consuming steps, the two reactions that produce ATP, and the two oxidation-reduction reactions that involve NAD^+-NADH. The three irreversible steps—those catalyzed by hexokinase, phosphofructokinase, and pyruvate kinase—control the rate of glycolysis. Of these three, phosphofructokinase is the most important enzyme controlling glycolysis.

Every glycolytic intermediate between glucose and pyruvate contains phosphate. Because phosphate is highly ionized, these intermediates cannot leave the cell. Besides trapping these compounds, the phosphate groups are also used to phosphorylate ADP to ATP.

The enzymes of glycolysis are in the cytosol rather than in the mitochondria. Since glycolysis is essential, the systemic lack of one of the eleven glycolytic enzymes is incompatible with life. The physician may, however, encounter patients whose erythrocytes and sometimes leukocytes are deficient in one or another glycolytic enzyme; they develop hemolytic anemia.

To initiate glycolysis, *hexokinase irreversibly phosphorylates glucose to glucose-6-P,* consuming ATP. As in all phosphorylation reactions, Mg^{+2} is an essential cofactor. The hexokinase reaction is one of the three rate-controlling steps of glycolysis. Hexokinase is inhibited by its endproduct, glucose-6-P. The liver possesses an additional kinase, glucokinase, with a higher Michaelis constant (K_m) than hexokinase to handle large surges of glucose after meals.

Glucose + ATP $\xrightarrow{\text{hexokinase}}$ Glucose-6-P + ADP

The second step of glycolysis is *the reversible isomerization of glucopyranose-6-P to fructofuranose-6-P by phosphoglucose isomerase.*
  Phosphofructokinase (PFK) then irreversibly phosphorylates fructose-6-P to fructose-1,6-bis-P. This rate-limiting step in glycolysis is under allosteric control (which is explained in the subsequent section on gluconeogenesis). Up to this point, for every mole of fructose-1,6-bis-P produced from glucose, two moles of ATP are consumed.

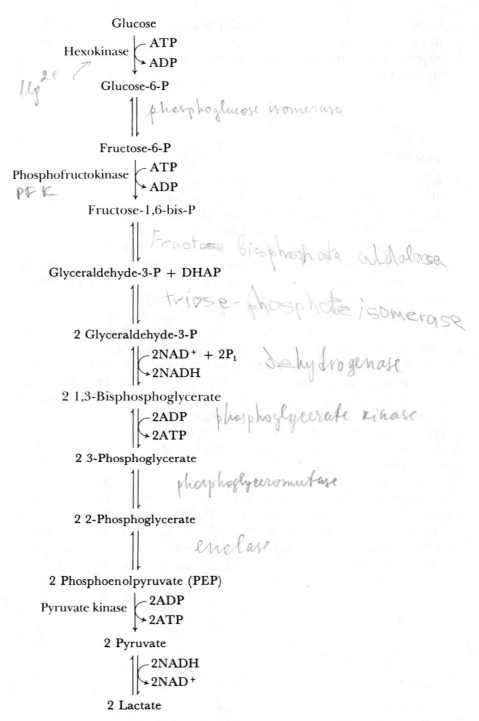

Fig. 8-1 Glycolysis. Single arrows denote reactions that are essentially irreversible, while double arrows represent reversible reactions.

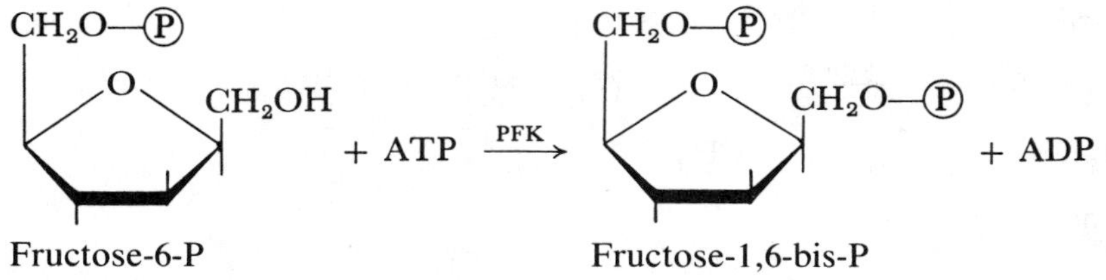

Fructose-6-P + ATP $\xrightarrow{\text{PFK}}$ Fructose-1,6-bis-P + ADP

Phosphofructokinase 2 (PFK2) differs from phosphofructokinase in that it phosphorylates fructose-6-P to fructose-2,6-bisphosphate. Fructose-2,6-bis-P stimulates PFK, thereby increasing glycolysis.

The next reaction marks the transition between the hexose and triose stages of glycolysis. *Fructose-bisphosphate aldolase reversibly cleaves the bond between carbons 3 and 4 of fructose-1,6-bis-P to create two triose phosphates: dihydroxyacetone phosphate (DHAP) and glyceraldehyde-3-P.*

DHAP can proceed no farther in glycolysis until its keto group is reversibly transformed into an aldehyde group by triose-phosphate isomerase to create another molecule of glyceraldehyde-3-P.

Finally, we arrive at the first energy-producing reaction of glycolysis, *the reversible glyceraldehyde-phosphate dehydrogenase reaction.* Two processes occur in this reaction:

1. The aldehyde group of glyceraldehyde-3-P is oxidized to a carboxylic acid group, thereby reducing NAD^+ to NADH.
2. Inorganic phosphate, P_i, joins to the OH group of this carboxyl group, creating 1,3-bisphosphoglycerate. Thus, ATP is not consumed in this phosphorylation.

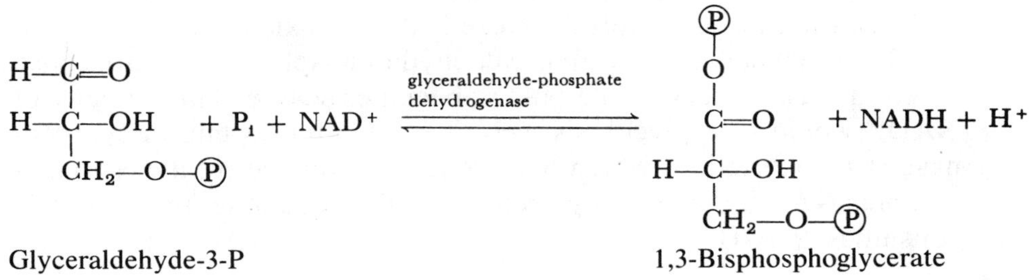

Glyceraldehyde-3-P + P_i + NAD^+ $\xrightleftharpoons{\text{glyceraldehyde-phosphate dehydrogenase}}$ 1,3-Bisphosphoglycerate + NADH + H^+

In erythrocytes, some of the 1,3-bisphosphoglycerate is converted to 2,3-bisphosphoglycerate (2,3-BPG); the latter shifts the O_2 dissociation curve of hemoglobin to the right (see Chap. 2).

The next step in glycolysis is *the removal of a phosphate from 1,3-bisphosphoglycerate by phosphoglycerate kinase to yield 3-phosphoglycerate plus ATP.*

Next, *phosphoglyceromutase shifts the phosphate from carbon 3 to carbon 2, yielding 2-phosphoglycerate:*

Enolase then dehydrates 2-phosphoglycerate to create phosphoenolpyruvate (PEP), an enol, i.e., a compound with the —C=C(OH)— structure:

$$\underset{\text{2-Phosphoglycerate}}{\begin{array}{c} \text{COO}^- \\ | \\ \text{H}-\text{C}-\text{O}-\textcircled{P} \\ | \\ \text{CH}_2\text{OH} \end{array}} \quad \underset{\text{enolase}}{\rightleftharpoons} \quad \underset{\text{PEP}}{\begin{array}{c} \text{COO}^- \\ | \\ \text{C}-\text{O}-\textcircled{P} \\ \| \\ \text{CH}_2 \end{array}} + \text{H}_2\text{O}$$

Because PEP has a very high-energy phosphate group (-14.8 kcal/mole), its hydrolysis easily drives ATP production, which requires 7.3 kcal/mole. *Pyruvate kinase catalyzes this irreversible phosphate transfer from PEP to ADP to yield pyruvate:*

$$\underset{\text{PEP}}{\begin{array}{c} \text{COO}^- \\ | \\ \text{C}-\text{O}-\textcircled{P} \\ \| \\ \text{CH}_2 \end{array}} + \text{ADP} \quad \xrightarrow{\underset{\text{kinase}}{\text{pyruvate}}} \quad \underset{\text{Pyruvate}}{\begin{array}{c} \text{COO}^- \\ | \\ \text{C}=\text{O} \\ | \\ \text{CH}_3 \end{array}} + \text{ATP}$$

Pyruvate kinase is an allosteric enzyme inhibited by ATP and alanine; it is activated by fructose-1,6-bisphosphate.

Pyruvate is a key link in metabolism that ties together glycolysis, the tricarboxylic acid (TCA) cycle, amino acid metabolism, and fatty acid oxidation.

In tissues with adequate oxygen supply, pyruvate formation is the last step of glycolysis. Most of this pyruvate is then oxidized and decarboxylated to form an acetyl group, which, when combined with coenzyme A (acetyl-CoA), enters the tricarboxylic acid cycle.

When the rate of glycolysis exceeds that of oxidative metabolism, NADH produced by the glyceraldehyde-phosphate dehydrogenase reaction accumulates, while NAD$^+$ becomes scarce. This scarcity of NAD$^+$ would stop glycolysis were it not for the lactate dehydrogenase (LDH) reaction, which uses NADH to reduce pyruvate to lactate plus NAD$^+$. **Anaerobic glycolysis,** therefore, neither produces nor consumes NADH.

$$\underset{\text{Pyruvate}}{\begin{array}{c} \text{COO}^- \\ | \\ \text{C}=\text{O} \\ | \\ \text{CH}_3 \end{array}} + \text{NADH} + \text{H}^+ \quad \underset{}{\overset{\text{LDH}}{\rightleftharpoons}} \quad \underset{\text{Lactate}}{\begin{array}{c} \text{COO}^- \\ | \\ \text{H}-\text{C}-\text{OH} \\ | \\ \text{CH}_3 \end{array}} + \text{NAD}^+$$

Lactic acid diffuses into the bloodstream and reaches the liver, where it is oxidized back to pyruvate and then metabolized aerobically.

In lactic acidosis, a common metabolic disorder, lactate is produced so rapidly that the liver cannot remove it fast enough. The blood lactic acid level rises, and metabolic acidosis ensues.

Energetics of Glycolysis

The overall reaction of aerobic glycolysis alone, using either free glucose, fructose, or galactose and yielding pyruvate, creates 2 moles of NADH and 4 moles of ATP but consumes 2 moles of ATP per mole of hexose; the net gain is 2 moles each of ATP and NADH.

Since glycolysis occurs only in the cytosol, the NADH produced must be transported into the mitochondria so that it can undergo oxidative phosphorylation. Since neither NADH nor NAD^+ can penetrate the mitochondrial membranes, shunts must transport this reducing power into the mitochondria. Two such shunts, discussed in Chapter 9, are the glycerol-phosphate and the malate shuttle, which yield 2 and 3 moles ATP, respectively, per mole of NADH transferred. Since there is still controversy about which shuttle dominates, the total net ATP gain in aerobic glycolysis is either 6 or 8 moles ATP per mole of hexose oxidized to pyruvate. The further oxidation of pyruvate to CO_2 and H_2O in the TCA cycle yields a total of 36 or 38 moles ATP per mole of hexose oxidized to CO_2 and H_2O.

In anaerobic glycolysis, on the other hand, only 2 moles of ATP and no NADH are produced per mole of hexose. Anaerobic tissues metabolize glucose much more rapidly than aerobic tissues do to compensate for this meager energy gain. The overall reaction of anaerobic glycolysis may thus be written:

$$\text{Glucose} + 2\text{ADP} + 2P_i \longrightarrow 2 \text{ lactate} + 2\text{ATP} + 2H_2O$$

Gluconeogenesis

Gluconeogenesis literally means the new formation of glucose; it involves the conversion of three- and four-carbon compounds to the six-carbon compound glucose. The fuels for gluconeogenesis may be carbohydrates such as glycerol and sugar alcohols or noncarbohydrates such as α-amino acids or α-keto acids like pyruvate and oxaloacetate.

The brain and exercising skeletal muscle require glucose as their principal fuel. During fasting, the liver stores only enough glycogen to supply the body with glucose for 12 to 24 hours. The sole source of glucose in prolonged fasting is that supplied by gluconeogenesis from glycerol (derived from triglyceride hydrolysis) and from α-keto acids (derived from amino acid catabolism); i.e., the glycerol backbone of triglycerides and proteins from enzymes and skeletal muscle are consumed to provide glucose for the brain and exercising skeletal muscle.

Gluconeogenesis is confined almost entirely to the liver, kidneys, and intestinal epithelium; certain enzymes required in gluconeogenesis are present only in these three organs.

The central pathway of gluconeogenesis from the α-keto acids is the conversion of pyruvate to glucose, a net reversal of glycolysis. Of the 11 reactions of glycolysis, 8 are reversible and can, therefore, be used in gluconeogenesis. There are 3 irreversible reactions of glycolysis, however, that must be bypassed in gluconeogenesis: the pyruvate-kinase, PFK, and hexokinase reactions.

Two enzymes are required to bypass the pyruvate-kinase reaction. **Pyruvate carboxylase,** using biotin as a coenzyme, carboxylates pyruvate to produce oxaloacetate (OAA). Almost all carboxylation of α-keto acids consume ATP and require biotin. Pyruvate carboxylase is an allosteric enzyme that requires the presence of acetyl-CoA. By supplying OAA, this pyruvate-carboxylase reaction is the principal means used to replenish the TCA cycle.

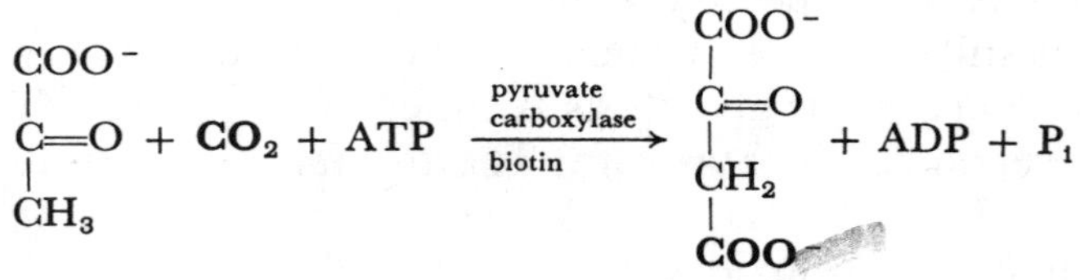

$$
\begin{array}{c}
\text{COO}^- \\
| \\
\text{C}{=}\text{O} \\
| \\
\text{CH}_3
\end{array}
+ \text{CO}_2 + \text{ATP}
\xrightarrow[\text{biotin}]{\text{pyruvate carboxylase}}
\begin{array}{c}
\text{COO}^- \\
| \\
\text{C}{=}\text{O} \\
| \\
\text{CH}_2 \\
| \\
\text{COO}^-
\end{array}
+ \text{ADP} + \text{P}_i
$$

Pyruvate   OAA   *phosphoenolpyruvate*

The second enzyme in this bypass is **PEP carboxykinase,** which phosphorylates and decarboxylates OAA to create PEP. This reaction requires GTP, which is obtained by phosphorylating GDP with ATP. As shown below, 1 mole of GTP has the same energy value as 1 mole of ATP:

guanosine diphosphate

$$\text{ATP} + \text{GDP} \rightleftharpoons \text{ADP} + \text{GTP}$$

Thus, 2 moles of ATP are consumed in bypassing the pyruvate-kinase reaction.

$$
\begin{array}{c}
\text{COO}^- \\
| \\
\text{C}{=}\text{O} \\
| \\
\text{CH}_2 \\
| \\
\text{COO}^-
\end{array}
+ \text{GTP}
\xrightarrow{\text{pyruvate carboxykinase}}
\begin{array}{c}
\text{COO}^- \\
| \\
\text{C}{-}\text{O}{-}\text{P} \\
\| \\
\text{CH}_2
\end{array}
+ \text{CO}_2 + \text{GDP}
$$

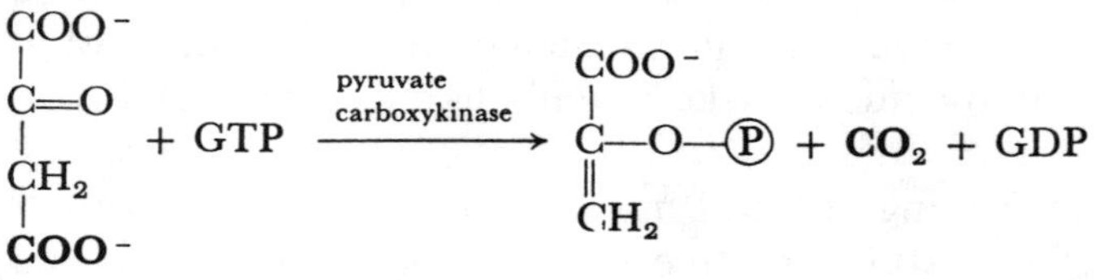

OAA   PEP

Once formed, PEP proceeds in the reverse direction through the steps of glycolysis until it reaches the next irreversible step: the PFK reaction. Gluconeogenesis uses fructose-1,6-bisphosphatase to remove the phosphate group from carbon 1 of fructose. This reaction does not regenerate the ATP consumed in the PFK reaction.

$$\text{Fructose-1,6-bis-P} + \text{H}_2\text{O} \xrightarrow{\text{fructose bisphosphatase}} \text{fructose-6-P} + \text{P}_i$$

If both the PFK and fructose bisphosphatase reactions occurred simultaneously, this would result in a futile cycle that consumed ATP.

Fortunately, both enzymes are allosterically controlled to prevent a futile cycle. When the cell has excess glucose, fructose-2,6-bis-P will be plentiful. Fructose-2,6-bis-P stimulates phosphofructokinase and pyruvate kinase, thereby increasing glycolysis; it inhibits fructose-1,6-bisphosphatase, which slows gluconeogenesis.

When the cell develops excess energy stores in the form of ATP, glycolysis is slowed because ATP inhibits phosphofructokinase and pyruvate kinase. Similarly, when the cell has excess citrate, gluconeogenesis is increased and glycolysis is slowed because citrate stimulates fructose-1,6-bisphosphatase and inhibits phosphofructokinase.

When the cell lacks glucose, fructose-2-6-bis-P levels fall. This inhibits glycolysis via phosphofructokinase and stimulates gluconeogenesis via the fructose bisphosphatase reaction. When the cell lacks energy, glycolysis increases and gluconeogenesis is slowed because AMP stimulates phosphofructokinase and inhibits fructose-1,6-bisphosphatase.

The next step of gluconeogenesis is the reversal of a glycolysis reaction: glucose-phosphate isomerase converts fructose-6-P to glucose-6-P.

Gluconeogenesis must then bypass the third irreversible step of glycolysis: the hexokinase reaction. **Glucose-6-phosphatase** removes phosphate from glucose-6-P to yield the endproduct, glucose. This reaction, which also yields inorganic phosphate, again fails to recover the ATP consumed in the hexokinase reaction:

$$\text{Glucose-6-P} + H_2O \xrightarrow{\text{glucose-6-phosphatase}} \text{glucose} + P_i$$

Type I glycogen storage disease results from the hereditary absence of glucose-6-phosphatase. Since this prevents the final step of gluconeogenesis, the liver and kidneys are unable to release glucose into the blood. Such individuals must eat every 3 to 5 hours to prevent fasting hypoglycemia.

The overall reaction for converting pyruvate to glucose can be summarized as follows:

$$2\text{ Pyruvate} + 2ATP + 2GTP \rightarrow \rightarrow 2PEP + 2ADP + 2GDP + 2P_i$$

$$2ATP + 2GDP \rightarrow \rightarrow 2ADP + 2GTP$$

$$2PEP + 2ATP \rightarrow \rightarrow \rightarrow 2\text{ 1,3-BPG} + 2ADP$$

$$2\text{ 1,3-BPG} + 2NADH + 2H^+ \rightarrow \rightarrow 2\text{ glyceraldehyde-3-P} + 2NAD^+ + 2P_i$$

$$2\text{ Glyceraldehyde-3-P} + 2H_2O \rightarrow \rightarrow \rightarrow \rightarrow \text{glucose} + 2P_i$$

Net: **2 Pyruvate + 6ATP + 2NADH + 2H$^+$ + 2H$_2$O $\longrightarrow$**

glucose + 6ADP + 6P$_i$ + 2NAD$^+$

Thus, gluconeogenesis from pyruvate consumes 6 moles ATP and 2 moles NADH. The reverse process—i.e., glycolysis to pyruvate—produces 2 moles each of ATP and NADH.

Gluconeogenesis from the glycerol (derived from triglyceride hydrolysis) begins by the phosphorylation of glycerol to glycerol-P and then the oxidation of this to DHAP. DHAP then proceeds in a reverse direction through the steps of glycolysis; hexosediphosphatase and glucose-6-phosphatase are again employed to bypass the two irreversible reactions of glycolysis.

An amino acid is termed **glucogenic** or **glycogenic** if it can be converted to glucose via gluconeogenesis. To enter gluconeogenesis, an amino acid must be broken down to pyruvate, 3-phosphoglycerate, or an intermediate in the tricarboxylic acid cycle. Sufficient dietary carbohydrate results in protein sparing, because amino acid breakdown for gluconeogenesis is reduced.

Pentose Phosphate Pathway

The **pentose phosphate pathway,** sometimes known as the **hexosemonophosphate shunt,** derives its name from the ribose phosphate produced. All reactions of this pathway occur in the cytosol. The main purposes of the pentose phosphate pathway are:

1. To produce ribose-5-P for nucleotide synthesis.
2. To produce NADPH from NADP$^+$ for fatty acid and steroid biosynthesis and for maintaining reduced glutathione inside erythrocytes.
3. To interconvert pentoses and hexoses.

Organs that actively synthesize fatty acids and steroids—such as the lactating mammary gland, the liver, the adrenal cortex, and adipose tissue—channel a significant proportion of their glucose into the pentose phosphate pathway. At least 30% of hepatic glucose enters this pathway.

In the initial reaction of the pathway, glucose-6-P dehydrogenase (G-6-PDH) oxidizes glucose-6-P to 6-phosphogluconolactone, producing NADPH:

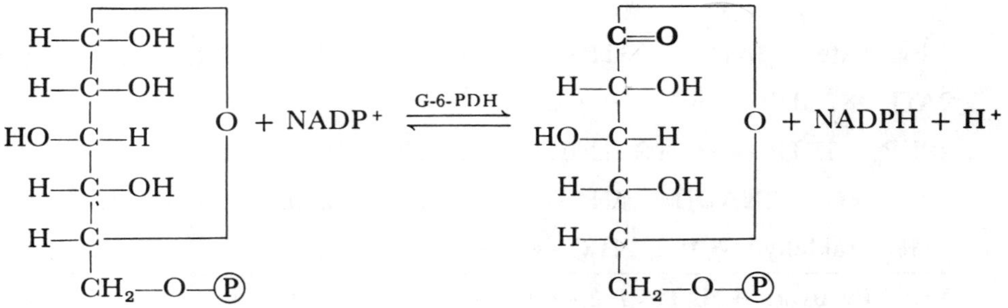

Glucose-6-P   6-Phosphogluconolactone

The inherited deficiency of glucose-6-P dehydrogenase in erythrocytes is one of the world's most common enzyme deficiency diseases, particularly among Mediterranean peoples and blacks. G-6-PDH deficiency stops the pentose phosphate pathway, thereby preventing

NADPH production. Erythrocytes need NADPH to maintain gluta-thione, a tripeptide, in its reduced form, to protect themselves from oxidizing agents. Individuals with this deficiency develop hemolytic anemia after receiving oxidizing drugs or eating fava beans (hence the name "favism" for this disease).

Subsequent reactions of the pathway decarboxylate the hexose sugars to form pentoses, such as ribose-5-P, an essential component of nucleic acids. Other reactions rearrange triose, tetrose, pentose, hexose, and heptulose sugars. The net overall reaction is:

$$6 \text{ Glucose-6-P} + 12\text{NADP}^+ + 7\text{H}_2\text{O} \longrightarrow$$
$$5 \text{ glucose-6-P} + 6\text{CO}_2 + 12\text{NADPH} + \text{P}_i + 12\text{H}^+$$

Because the NADPH from this pathway can be regenerated anaero-bically to NADP^+ in fatty acid and steroid synthesis, the pathway can operate anaerobically.

In this pathway, there is a transketolase reaction that requires TPP as its coenzyme. In thiamine deficiency, the transketolase activity in erythrocytes is reduced.

Glycogenolysis

Glycogenolysis means the lysis or breakdown of glycogen. The first step of glycogen breakdown is the phosphorylation of its α-1,4 gly-cosidic bonds by glycogen phosphorylase. Glycogen phosphorylase exists in two forms: phosphorylase b, which is usually inactive, and phosphorylase a, which is usually active. Both phosphorylases in turn exist in two forms: R (relaxed) and T (tense). Phosphorylase kinase and ATP are required to convert phosphorylase b to phos-phorylase a:

Phosphorylase b + ATP ⟶ Phosphorylase a + ADP

Phosphorylase b is an allosteric enzyme activated by high levels of AMP; ATP and glucose-6-P are allosteric inhibitors. In most cases phosphorylase b remains inactive because the allosteric effects of ATP and glucose-6-P outweigh those of AMP.

Most of the time phosphorylase a is in the active R form. High concentrations of glucose, however, convert it to an inactive T form by allosteric inhibition.

Figure 8-2 shows how glycogen breakdown and synthesis are recip-rocally regulated. In skeletal muscle, epinephrine activates adenylate cyclase but glucagon does not. In the liver, glucagon is the principal activator of adenylate cyclase; epinephrine, however, can also acti-vate this enzyme. Epinephrine and glucagon bind to receptors on the cell membrane, thereby activating adenylate cyclase. Adenylate cy-clase converts ATP to 3',5'-cyclic AMP. Cyclic AMP can be inacti-vated by phosphodiesterase. If it remains active, it allosterically ac-

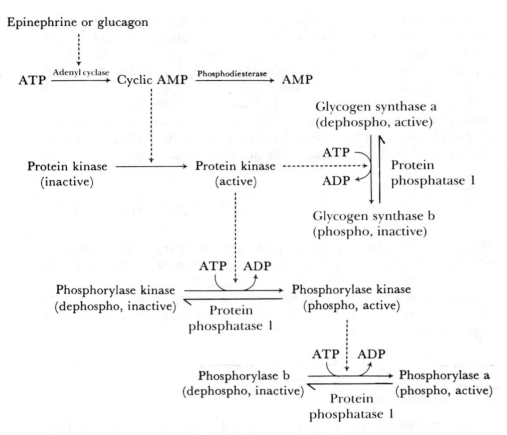

Fig. 8-2 Cyclic AMP-mediated activation of glycogen phosphorylase and inhibition of glycogen synthase.

tivates protein kinase. Muscle protein kinase is partially activated by a second mechanism, high calcium levels, but liver protein kinase is not. Calcium released during muscle contraction binds to calmodulin, a calcium-binding polypeptide that is a component of muscle protein kinase. This binding partially activates protein kinase. Hence calcium binding to calmodulin helps muscles obtain glucose from glycogen during exercise.

The active protein kinase phosphorylates two enzymes: phosphorylase kinase and glycogen synthase (see Fig. 8-2). When phosphorylated, phosphorylase kinase becomes active, whereas glycogen synthase becomes inactive. Thus, protein kinase reciprocally turns glycogen breakdown on while turning glycogen synthesis off.

The active phosphorylase kinase promotes the conversion of phosphorylase b to phosphorylase a, the active form. Phosphorylase a then proceeds to cleave the 1,4 linkages of glycogen.

Protein phosphatase 1 also assists in the reciprocal control of glycogen breakdown and synthesis. Protein phosphatase 1 removes the phosphate from phosphorylase kinase and phosphorylase a, thereby slowing glycogen breakdown. In contrast, protein phosphatase 1 removes the phosphate from inactive glycogen synthase b, converting it to the active glycogen synthase a (see Fig. 8-2).

In addition to activation by covalent phosphate binding, both glycogen phosphorylase and glycogen synthase are also allosterically regulated. AMP allosterically stimulates glycogen phosphorylase b, and glucose-6-P stimulates glycogen synthase. Glucose allosterically inhibits glycogen phosphorylase.

Cyclic-AMP phosphodiesterase inactivates cyclic AMP. Theophylline acts to dilate bronchi by inhibiting cyclic-AMP phosphodiesterase.

In the process of glycogenolysis, glycogen phosphorylase a yields a limit dextrin after it has released successive glucose-1-P residues from the branch chains of the glycogen. First, oligo-1,4→1,4-glucan transferase shifts oligosaccharides so that a single 1,6-bound glucose remains at the branch point. Next, amylo-1,6-glucosidase cleaves the α-1,6 linkages.

The hereditary absence of muscle glycogen phosphorylase, termed type V glycogen storage disease or McArdle's disease, causes normal glycogen to accumulate in skeletal muscle. During vigorous exercise glycogen breakdown in muscle is impaired, causing muscle cramps and limited exercise tolerance.

Type III glycogen storage disease, the inherited deficiency of amylo-1,6-glucosidase or oligo-1,4→1,4-glucan transferase glucosidase in the liver and in heart and skeletal muscle, causes limit dextrins to accumulate in these tissues.

Phosphoglucomutase converts the glucose-1-P liberated in glycogenolysis to glucose-6-P, which may have one of three fates: it can proceed through glycolysis, it can enter the pentose phosphate pathway, or, in the liver, kidneys, and intestinal epithelium, it can be cleaved by glucose-6-phosphatase to form free glucose and inorganic phosphate.

Glycogenesis

Glycogenesis means glycogen synthesis. The initial step of glycogenesis is the conversion of glucose-6-P to glucose-1-P by phosphoglucomutase. Next, glucose-1-P reacts with UTP to create UDP-glucose. This reaction resembles the binding of galactose to UTP to form UDP-galactose.

$$\text{Glucose-1-P} + \text{UTP} \longrightarrow \text{UDP-glucose} + PP_i$$

Glycogen synthase then adds this bound glucose in α-1,4 linkage to the glycogen polymer, liberating UDP:

$$\text{UDP-glucose} + (\text{glucose})_n \longrightarrow (\text{glucose})_{n+1} + \text{UDP}$$

For every glucose molecule incorporated into glycogen in glycogenesis, 1 mole of ATP is expended to produce glucose-6-P from glucose, and 1 mole of UTP is spent to create UDP-glucose. Following glycogenolysis, the 1 mole of ATP spent in glycogen synthesis is recovered when the glucose-1-P produced undergoes glycolysis; in fact, 3 moles of ATP are generated by glycolysis per mole of glucose-1-P removed from glycogen.

In glycogenesis, an α-1,4 glucan branching enzyme removes α-1,4-linked glucose oligosaccharides and reattaches them by α-1,6 bonds to create the proper branching. In type IV glycogen storage disease, the hereditary absence of this branching enzyme leads to the accumulation of long glucose polymers with few branches.

The process of glycogenesis is coupled somehow to the influx of K^+ into the cells. Hyperkalemia (a high serum K^+ level) is usually treated initially by giving glucose and insulin to induce glycogenesis and the concomitant removal of K^+ from the serum.

Hormonal Control of Carbohydrate Metabolism

Glucose does not readily penetrate through the cell membranes of most human tissues. To facilitate glucose entry into cells, insulin must be present. The brain, liver, kidneys, and blood cells, however, do not need insulin for glucose transport. The rapid intravenous injection of 50 units of regular insulin, for instance, will lower the serum glucose level rapidly and cause insulin shock, which is characterized by impaired consciousness or coma, sweating, anxiety, and various neurologic abnormalities that result from an inadequate glucose supply to the brain. Hypoglycemic shock is treated by administering intravenous glucose.

Insulin also promotes amino acid uptake by cells and stimulates protein synthesis, thereby reducing the amino acid supply available for gluconeogenesis. By inhibiting the synthesis of the key enzymes of gluconeogenesis, insulin again helps to lower the serum glucose level.

In addition, insulin promotes glycogenesis by stimulating glycogen synthase. Insulin not only lowers cyclic AMP but also makes protein kinase less sensitive to cyclic AMP, thereby blunting the glycogenolytic response to epinephrine or glucagon. Insulin is the only hormone that acts to lower serum glucose levels as well as to promote glucose storage.

Somatotrophic hormone (STH), also termed growth hormone (GH), antagonizes insulin activity by an unknown mechanism. Thus, STH is a glucose-mobilizing hormone; it raises the serum glucose level. Hyperglycemia is a feature of acromegaly, which is caused by an STH-producing pituitary tumor.

Somatomedins are proteins that have a very similar structure to proinsulin; their hormonal effects resemble those of insulin. They promote glucose and amino acid uptake by cells and stimulate the synthesis of glycogen, protein, and triglyceride. Their most characteristic action is to promote the sulfation of glycoproteins.

Thyroid hormones also have some effect in raising serum glucose levels. The mechanism for this is uncertain.

Glucocorticoids (the 11-hydroxy, C21 adrenocortical steroids) promote hyperglycemia by inhibiting glycolysis and stimulating gluconeogenesis. They promote protein and amino acid breakdown, thereby providing more pyruvate and OAA as fuel for gluconeogenesis. In addition, they stimulate the liver to produce more gluconeogenetic enzymes, such as glucose-6-phosphatase, pyruvate carboxylase, and PEP carboxykinase.

Epinephrine and glucagon are also glucose-mobilizing hormones. They raise the serum glucose level mainly by stimulating glycogenolysis via cyclic AMP in the liver and kidneys, as well as by inhibiting glycogenesis (see Fig. 8-2).

Diabetes mellitus has traditionally been considered as a deficiency of effective insulin action. More recently, however, it has been suggested that the inappropriately high serum glucagon levels in diabetes mellitus contribute to the hyperglycemia of diabetes mellitus.

The increase in serum glucose levels after a carbohydrate meal triggers insulin release while inhibiting the glucose-mobilizing hormones such as epinephrine, glucagon, and the glucocorticoids. Several hours after a meal, insulin secretion diminishes, and the levels of the glucose-mobilizing hormones increase. During fasting, virtually no insulin is detectable in the serum. Glucagon and epinephrine can maintain the serum glucose concentration for only 6 to 24 hours. After the hepatic glycogen reserves are depleted, glucocorticoids assume the dominant role of providing glucose via gluconeogenesis.

Problems

Problem 1

With respect to fructose and galactose metabolism:

    A. The initial step is conversion to sugar-6-phosphate in the liver.
    B. Fructose and galactose are both metabolized to UDP-glucose.
    C. Deficiency of the kinase enzymes for fructose and galactose does not cause disease.
    D. Fructose but not galactose can be converted to glycogen.
    √E. Hexokinase phosphorylates fructose but not galactose.

Problem 2

Which of the following mechanisms does not govern the activity of phosphofructokinase (PFK)?

    A. Fructose-2,6-bisphosphate stimulates PFK.
    B. ATP inhibits PFK.
    √C. Fructose-1,6-bisphosphate stimulates PFK.
    D. AMP stimulates PFK.

Problems 3–6
Gluconeogenesis must bypass three irreversible reactions of glycolysis. Match the following to the enzymes of gluconeogenesis given in Problems 3–6.

A. Enolase
B. Pyruvate kinase
C. Hexokinase
D. PFK

3. Fructose bisphosphatase.
4. Glucose-6-phosphatase.
5. Pyruvate carboxykinase.
6. Pyruvate carboxylase.

Problem 7
Which of the following statements about fructose-2,6-bisphosphate is incorrect?

A. Stimulates fructose-1,6-bisphosphatase.
B. Formed from fructose-6-P.
C. Formed by PFK2.
D. Stimulates PFK.
E. Degraded to fructose-1-phosphate by fructose-1,6-bisphosphatase.

Problems 8–10
Match the statements in Problems 8–10 to the choices below.

A. Protein kinase
B. Phosphorylase kinase
C. Phosphorylase b
D. Phosphorylase a

8. Cleaves 1,4-glycosidic linkages in glycogen.
9. Activated by cyclic AMP.
10. Inactivates glycogen synthase.

Problems 11–13
Match the following enzyme deficiencies to the conditions listed in Problems 11–13.

A. Glycogen phosphorylase deficiency
B. Glucose-6-phosphatase deficiency
C. Amylo-1,6-glucosidase deficiency
D. Glucose-6-P dehydrogenase deficiency

11. Glycogen with abnormal branches accumulates.
12. Blocks the pentose phosphate pathway in erythrocytes.
13. Hypoglycemia (low serum glucose concentration).

Problems 14–18
Match the following hormones to their actions on carbohydrate metabolism, given in Problems 14–18.

A. Insulin
B. Glucocorticoids
C. Somatotrophic hormone
D. Glucagon
E. Epinephrine
F. Thyroid hormones

14. Hypoglycemic effect.
15. Potent stimulant of gluconeogenesis.
16. Inhibits gluconeogenesis.
17. Produced in the pancreatic islet cells, this hormone causes rapid glycogenolysis to counteract hypoglycemia.
18. Stimulates glycogen synthase.

Answers

1. E.
2. C.
3. D.
4. C.
5. B.
6. B.
7. A. It inhibits this enzyme.
8. D.
9. A.
10. A.
11. C.
12. D.
13. B.
14. A.
15. B.
16. A.
17. D.
18. A.

References

Devlin, T. M. *Textbook of Biochemistry with Clinical Correlations* (3rd ed.). New York: Wiley-Liss, 1992. Pp. 292–386.

Mathews, C. K., and van Holde, K. E. *Biochemistry.* Redwood City, Calif.: Benjamin/Cummings, 1990. Pp. 433–466, 493–501, 538–556.

Murray, R. K., Granner, D. K., Mayes, P. A., and Rodwell, V. W. *Harper's Biochemistry* (22nd ed.). Norwalk, Conn.: Appleton & Lange, 1990. Pp. 163–198.

Styer, L. *Biochemistry* (3rd ed.). New York: Freeman, 1988. Pp. 349–372, 427–468.

Tricarboxylic Acid Cycle and Oxidative Phosphorylation

Unlike carbohydrate metabolism, which takes place in the cytoplasm, the tricarboxylic acid (TCA) cycle and oxidative phosphorylation occur within the mitochondria.

The Tricarboxylic Acid Cycle

The **tricarboxylic acid cycle,** also called the citric acid or Krebs cycle, is so named because several of its intermediates have three carboxyl groups: citrate, *cis*-aconitate, and isocitrate. The remaining six intermediates are dicarboxylic acids.

The TCA cycle is the central hub in the metabolism of carbohydrates, fatty acids, and amino acids. Although its primary function is energy production, it also provides intermediates for synthesizing the amino acids and porphyrins.

Erythrocytes differ from most other human cells in that they lack mitochondria, and cannot therefore use the TCA cycle.

The TCA cycle oxidizes acetic acid completely to 2 moles CO_2 plus eight hydrogen atoms, which enter the electron-transport chain:

$$CH_3COOH + 2H_2O \longrightarrow 2CO_2 + 8H$$

Several sources provide the acetate to fuel the TCA cycle. The main source of acetate from glycolysis is the mitochondrial **pyruvate-dehydrogenase reaction.** Pyruvate formed in glycolysis or in the catabolism of five of the amino acids readily penetrates into the mitochondria. Thiamine pyrophosphate (TPP) decarboxylates pyruvate to CH_3CHOH-TPP. Oxidized lipoic acid, a second coenzyme, oxidizes this CH_3CHOH— to $CH_3C(=O)$— and transfers it to CoA-SH, yielding acetyl-CoA and reduced lipoic acid. FAD returns lipoic acid to its oxidized state, generating $FADH_2$, which in turn reduces NAD^+ to NADH. Thus, five coenzymes participate in the pyruvate-dehydrogenase reaction (see illustration in Chap. 6, under Thiamine Pyrophosphate). The overall reaction is:

$$Pyruvate + CoA\text{-}SH + NAD^+ \longrightarrow acetyl\text{-}CoA + CO_2 + NADH + H^+$$

Two enzymes regulate pyruvate dehydrogenase (PDH). PDH kinase, stimulated by high ratios of ATP/ADP, NADH/NAD, or acetyl-CoA/CoA, phosphorylates the active form of PDH, making it inactive. PDH phosphatase, stimulated by insulin and calcium, removes the phosphate from inactive PDH, thereby reactivating it.

The endproducts of PDH dehydrogenase, acetyl-CoA and NADH, inhibit PDH, whereas AMP stimulates the enzyme complex.

β-Oxidation of fatty acid, occurring within the mitochondria, is a second source of acetyl-CoA for the TCA cycle, as is the degradation of ketone bodies. Another source of acetyl-CoA is from the catabolism of eight of the amino acids.

As shown below and in Figure 9-1, **citrate synthase** brings acetate from acetyl-CoA into the TCA cycle by joining its methyl group to the keto group of oxaloacetate (OAA) to produce citrate, a tricarbox-

Fig. 9-1   The tricarboxylic acid (TCA) cycle.

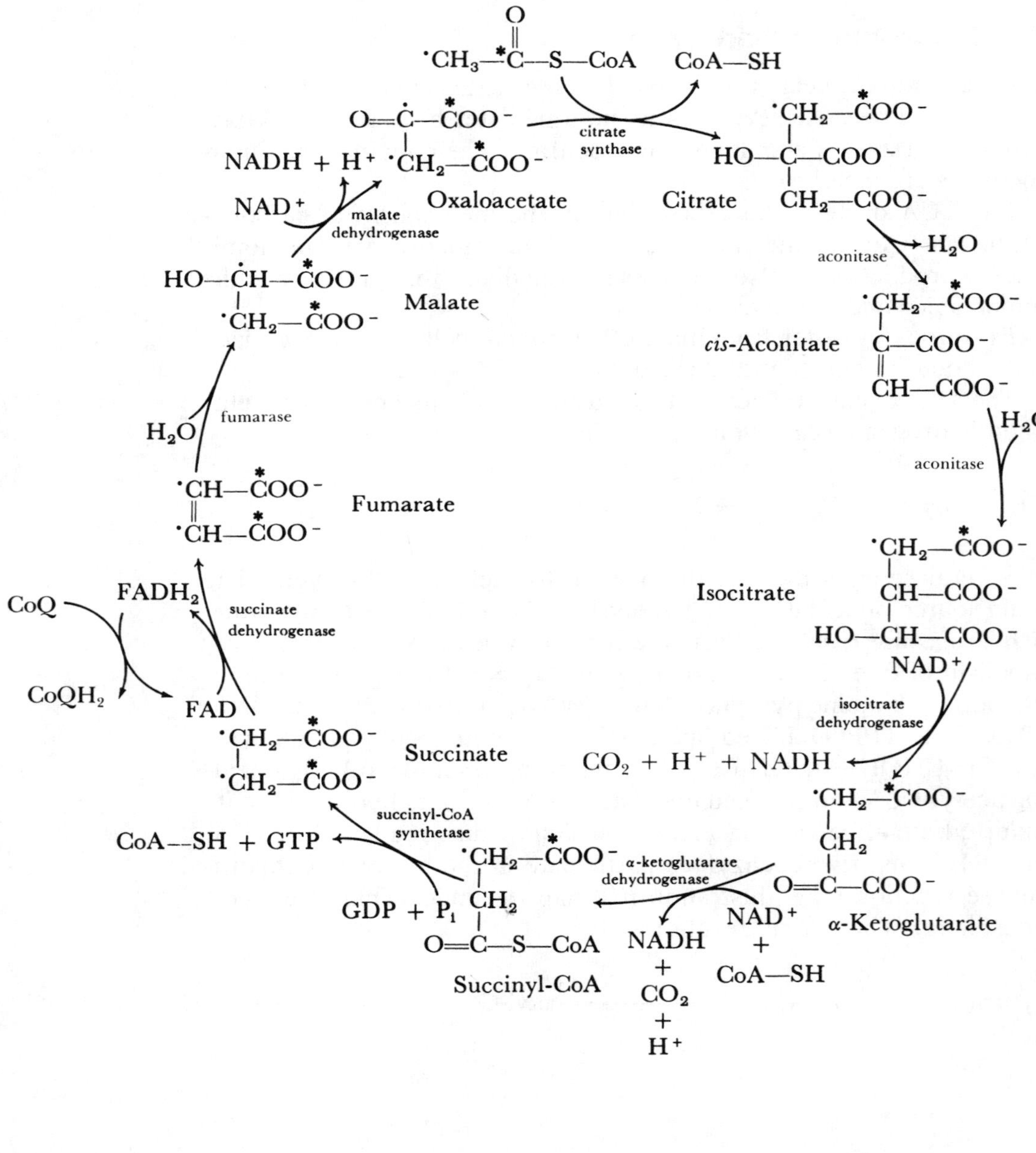

ylic acid. In Figure 9-1 and in the reactions to follow, the asterisk and dot markers are used to trace the fate of the carbon atoms of the carboxyl and methyl groups of the initial acetate:

$$\cdot CH_3 - {}^*\overset{\displaystyle O}{\overset{\|}{C}} - S - CoA \quad + \quad \underset{\displaystyle CH_2 - COO^-}{\overset{\displaystyle O=C-COO^-}{|}} \quad \xrightarrow[\text{synthase}]{\text{citrate}} \quad \underset{\displaystyle CH_2 - COO^-}{\overset{\displaystyle \cdot CH_2 - {}^*COO^-}{\underset{|}{HO-C-COO^-}}} \quad + \quad CoA\text{-}SH$$

Acetyl-CoA OAA Citrate

Because citrate synthase catalyzes the first committed step of the TCA cycle, it is not surprising that citrate synthase helps control the rate of the cycle. ATP allosterically inhibits citrate synthase. The availability of oxaloacetate and acetyl-CoA is the most important factor regulating citrate synthase.

Aconitase dehydrates citrate to create a carbon-carbon double bond in *cis*-aconitate (a hydratase adds water to a double bond or removes water to create a double bond). This hydratase then rehydrates *cis*-aconitate to yield isocitrate. The overall effect of these two reactions is to move the hydroxyl group of citrate to another carbon atom to create isocitrate.

Although citrate appears to be symmetrical, the two CH_2COOH ends differ with respect to spatial orientation of the OH and COOH groups. Aconitase can distinguish between the two ends of the molecule groups. Hence, if the carbons of acetate are radioactively labeled, the labels will remain fixed in relative position, rather than appearing at both ends of the isocitrate molecules.

$$\underset{\displaystyle CH_2 - COO^-}{\overset{\displaystyle \cdot CH_2 - {}^*COO^-}{\underset{|}{\mathbf{HO}-C-COO^-}}} \quad \underset{\displaystyle H_2O}{\overset{\displaystyle H_2O}{\rightleftharpoons}} \quad \underset{\displaystyle CH-COO^-}{\overset{\displaystyle \cdot CH_2 - {}^*COO^-}{\underset{\|}{C-COO^-}}} \quad \underset{\displaystyle H_2O}{\overset{\displaystyle H_2O}{\rightleftharpoons}} \quad \underset{\displaystyle \mathbf{HO}-CH-COO^-}{\overset{\displaystyle \cdot CH_2 - {}^*COO^-}{\underset{|}{CH-COO^-}}}$$

Citrate *cis*-Aconitate Isocitrate

Isocitrate dehydrogenase oxidizes the hydroxyl group of isocitrate to an α-keto group and removes the middle carboxyl group, yielding α-ketoglutarate (a five-carbon dicarboxylic acid), CO_2, and NADH. This NAD^+-linked isocitrate-dehydrogenase reaction is a rate-controlling step in the TCA cycle. Its positive allosteric modifier is ADP, while its negative modifiers are NADH and ATP. An $NADP^+$-linked isocitrate dehydrogenase also exists in the cytoplasm to create NADPH for fatty-acid and steroid biosynthesis. This enzyme does not participate in the TCA cycle.

$$\underset{\displaystyle \mathbf{HO}-CH-COO^-}{\overset{\displaystyle \cdot CH_2 - {}^*COO^-}{\underset{|}{CH-COO^-}}} \quad + NAD^+ \quad \underset{\text{dehydrogenase}}{\overset{\text{isocitrate}}{\rightleftharpoons}} \quad \underset{\displaystyle O\overset{\alpha}{=}C-COO^-}{\overset{\displaystyle \gamma\cdot CH_2 - {}^*COO^-}{\underset{|}{\beta CH_2}}} \quad + CO_2 + NADH + H^+$$

Isocitrate α-Ketoglutarate

In the liver, α-ketoglutarate is transaminated to produce glutamate; conversely, the α-amino group is removed from glutamate to yield α-ketoglutarate. Thus, the pool of α-ketoglutarate fluctuates.

The **α-ketoglutarate-dehydrogenase** complex resembles that of pyruvate dehydrogenase; both decarboxylate and dehydrogenate a CoA-bound keto acid and produce NADH, and both require lipoic acid and thiamine pyrophosphate. α-Ketoglutarate dehydrogenase removes the α-carboxyl group of α-ketoglutarate to yield, in the presence of CoA-SH, succinyl-CoA. This enzyme is allosterically inhibited by its endproducts, succinyl-CoA and NADH, and by ATP.

$$\begin{array}{l} \dot{C}H_2-*COO^- \\ | \\ CH_2 \\ | \\ O=C-\mathbf{COO}^- \end{array} + CoA\text{-}SH + NAD^+ \rightleftharpoons \begin{array}{l} \dot{C}H_2-*COO^- \\ | \\ CH_2 \\ | \\ O=C-S-CoA \end{array} + \mathbf{CO_2} + NADH + H^+$$

α-Ketoglutarate   Succinyl-CoA

In the liver, succinyl-CoA is siphoned from the TCA cycle for porphyrin synthesis, but replaced by means of the degradation of certain amino acids to succinyl-CoA.

Succinyl-CoA synthetase removes CoA from succinyl-CoA to liberate succinate. This enzyme is named for the reverse reaction, the synthesis of succinyl-CoA. This cleavage of CoA drives the substrate-level phosphorylation of GDP to GTP; it is the only substrate-level phosphorylation in the TCA cycle. **Substrate-level phosphorylations** such as those in glycolysis, e.g., the conversion of PEP + ADP to pyruvate + ATP, were discussed in Chapter 8. Oxidative phosphorylation, in contrast, occurs by means of electron transport through the cytochrome system.

$$\begin{array}{l} \dot{C}H_2-*COO^- \\ | \\ CH_2 \\ | \\ O=C-S-CoA \end{array} + GDP + P_i \rightleftharpoons \begin{array}{l} \dot{C}H_2-*COO^- \\ | \\ \dot{C}H_2-*COO^- \end{array} + GTP$$

Succinyl-CoA   Succinate

The symmetry of succinate disperses the labels from the initial acetate molecule to both ends of the succinate molecules. Thus, both carboxyl groups of succinate are marked as being derived from the carboxyl group of acetate. Similarly, both —CH_2— groups are tagged as coming from the methyl group of acetate. The two CO_2 molecules liberated on the first turn of the cycle do not come from the initial acetate, but subsequent turns will liberate labeled CO_2 from the initial labeled acetate.

The GTP produced in the succinyl-CoA-synthetase reaction then phosphorylates ADP to ATP:

$$GTP + ADP \rightleftharpoons GDP + ATP$$

Succinate dehydrogenase oxidizes succinate with FAD to create the —CH=CH— bond of fumarate. Succinate dehydrogenase is the only FAD-linked dehydrogenase in the TCA cycle; isocitrate dehydrogenase, α-ketoglutarate dehydrogenase, and malate dehydrogenase are NAD$^+$-linked.

$$\begin{array}{l} \cdot CH_2-*COO^- \\ | \\ \cdot CH_2-*COO^- \end{array} + FAD \xrightarrow{\text{succinate dehydrogenase}} \begin{array}{l} \cdot CH-*COO^- \\ \| \\ \cdot CH-*COO^- \end{array} + FADH_2$$

Succinate   Fumarate

Fumarase then hydrates the —HC=CH— bond of fumarate to create malate:

$$\begin{array}{l} \cdot CH-*COO^- \\ \| \\ \cdot CH-*COO^- \end{array} + H_2O \xrightleftharpoons{\text{fumarase}} \begin{array}{l} HO-\cdot CH-*COO^- \\ | \\ \cdot CH_2-*COO^- \end{array}$$

Fumarate   Malate

To complete the cycle, **malate dehydrogenase** uses NAD$^+$ to oxidize the hydroxyl group of malate to the keto group of oxaloacetate:

$$\begin{array}{l} HO-\cdot CH-*COO^- \\ | \\ \cdot CH_2-*COO^- \end{array} + NAD^+ \xrightarrow{\text{malate dehydrogenase}} \begin{array}{l} O=\cdot C-*COO^- \\ | \\ \cdot CH_2-*COO^- \end{array} + NADH + H^+$$

Malate   OAA

The TCA cycle is controlled at three steps. The first control point is the citrate-synthase reaction, inhibited by ATP. The next control point is the isocitrate-dehydrogenase reaction, also inhibited by ATP; ADP stimulates this enzyme. The final control point is α-ketoglutarate dehydrogenase, which is inhibited by its endproducts, succinyl-CoA and NADH. Overall, the TCA cycle is inhibited by high ATP levels. If oxidative phosphorylation ceases, the TCA cycle will be shut down due to lack of NAD$^+$ and FAD.

The TCA cycle yields 3 moles NADH, 1 mole FADH$_2$, and 1 mole GTP for every acetate unit it oxidizes.

TCA intermediates such as α-ketoglutarate and OAA are often removed from the cycle to serve as precursors for glutamate and aspartate, respectively. To replenish these intermediates, pyruvate carboxylase synthesizes OAA from pyruvate. In carbohydrate deficiency, there is not enough pyruvate for the pyruvate-carboxylase reaction, which causes depletion of the TCA-cycle intermediates and excess fat mobilization.

Oxidative Phosphorylation
Porphyrins, Heme- and Non-Heme Iron-Sulfur Proteins

The players in the oxidative-phosphorylation game consist of heme-proteins as well as non-heme, iron-containing proteins.

The fundamental unit in the structure of heme is the pyrrole ring, shown below. Four pyrrole rings are linked to one another by $=CH—$ bridges to create porphin (shown below), the parent compound to the porphyrins. Side chains are added to the porphin ring to yield proto-porphyrin. Ferrous iron (Fe^{+2}) then binds to the central four nitrogen atoms of protoporphyrin to yield heme:

Pyrrole   Porphin   Iron binding in heme

Heme, when conjugated with a protein, forms a **heme-protein,** or hemo-protein; the major heme-proteins are hemoglobin, an O_2 carrier, and the cytochromes, the electron carriers. Cytochrome P_{450}, which is found in liver microsomes, differs from other cytochromes in that it does not participate in oxidative phosphorylation; it hydroxylates steroids and drugs.

There are two hydrogen-electron carriers in oxidative phosphorylation that contain non-heme, iron-sulfur proteins. Their iron atom binds directly to the enzyme without requiring a porphyrin ring. These carriers are succinate dehydrogenase and NADH dehydrogenase.

Humans synthesize porphyrins and heme from glycine and the TCA intermediate, succinyl-CoA. Aminolevulinic acid (ALA) synthetase uses pyridoxal phosphate as a coenzyme to join glycine with succinyl-CoA to yield ALA. Vitamin B_6 deficiency produces hypochromic microcytic anemia, which is the result of inadequate hemoglobin synthesis, by depriving ALA synthetase of its coenzyme.

The ALA synthetase reaction controls the rate of porphyrin synthesis. Heme, the endproduct of porphyrin synthesis, both inhibits and represses synthesis of ALA synthetase. ALA is converted to porphobilinogen, a compound consisting of a pyrrole nucleus with attached side chains. Four molecules of porphobilinogen then join to form an uroporphyrin. Uroporphyrins are decarboxylated to coproporphyrins, which are then oxidized to protoporphyrins. Heme synthetase adds Fe^{+2} to protoporphyrin to yield heme.

The **porphyrias** are a class of acquired or inherited disorders characterized by increased serum levels and urinary excretion of porphyrins. When exposed to sunlight, urine specimens from patients with such disorders will often darken.

Lead poisoning, an acquired porphyria, inhibits several enzymes of porphyrin synthesis, particularly heme synthetase. The inhibition of heme synthetase results in hypochromic microcytic anemia and the overproduction of coproporphyrins and ALA. Attacks of acute abdominal pain and neurologic impairment also ensue.

The inherited porphyrias usually stem from an overactive ALA synthetase enzyme that is not inhibited by heme. Except for acute intermittent porphyria, the inherited porphyrias cause photosensitivity dermatitis, possibly due to fluorescence of the excess porphyrins. Acute intermittent porphyria causes attacks of abdominal pain similar to those in lead poisoning.

The Electron Transport Chain

The central core of electron transport consists of three protein complexes transferring electrons from NADH to oxygen: NADH-Q reductase, cytochrome reductase, and cytochrome oxidase.

The NADH-Q reductase complex is an enzyme with CoQ (ubiquinone), FMN, and Fe-S prosthetic groups. Electron transport proceeds in these steps:

1. Two electrons from NADH are initially transferred to FMN, creating $FMNH_2$.
2. $FMNH_2$ then transfers its electrons to the Fe-S prosthetic groups, reducing iron from Fe^{+3} to Fe^{+2}.
3. Electrons from Fe^{+2} are then transferred to ubiquinone, reducing it to ubiquinol.

The second player in electron transport is the cytochrome reductase complex, which consists of cytochromes b and c and other proteins and has CoQ, heme, and Fe-S as prosthetic groups. Electron transport through cytochrome reductase proceeds in these steps:

1. Ubiquinol transfers an electron to the Fe-S complex, reducing Fe^{+3} to Fe^{+2}. In the process ubiquinol (QH_2) is converted to the semiquinone ($QH\cdot$).
2. The Fe-S complex then transfers its electron to cytochrome c_1, which passes it on to cytochrome c.
3. Cytochrome b then catalyzes the transfer of electrons between two semiquinone molecules, forming ubiquinone plus ubiquinol.

The third player in electron transport is the cytochrome oxidase complex, which consists of cytochromes a and a_3 with two different heme and copper prosthetic groups. Heme a is located near Cu_A, while heme a_3 is located near Cu_B. Electron transport through cytochrome oxidase occurs in these steps:

1. Electrons from cytochrome c are transferred to the heme a-Cu_A complex.

2. The heme a-Cu_A complex then transfers its electron to the heme a_3-Cu_B complex, reducing Fe^{+3} and Cu^{+2} to Fe^{+2} and Cu^+ respectively.

3. The Fe^{+2} and Cu^+ each donate an electron to O_2, converting it in the presence of H^+ to H_2O.

Complete reduction of oxygen yields water. Partial reduction, however, produces dangerous compounds such as superoxide anions (O_2^-), hydroxyl radicals (OH·), and hydrogen peroxide (H_2O_2). Human cells use superoxide dismutases to convert superoxide to hydrogen peroxide:

$$2O_2^- + 2H^+ \longrightarrow H_2O_2 + O_2$$

Catalase then converts hydrogen peroxide to water:

$$H_2O_2 \longrightarrow H_2O + \tfrac{1}{2}O_2$$

Peroxidases can also reduce hydrogen peroxide to water by using various reducing agents.

Breathing 100% oxygen for prolonged periods is harmful in large part because of the toxic oxygen radicals generated. Antioxidants such as α-tocopherol, β-carotene, ascorbate, and glutathione inactivate these radicals.

Neutrophilic leukocytes produce both superoxide anion and hydrogen peroxide to kill bacteria. Impaired production of these toxic intermediates in chronic granulomatous disease, a genetic disorder, leads to recurring infections of the sinuses and lungs.

NADH is not the only entry point to electron transport. Three other pathways reduce ubiquinone to ubiquinol. None, however, act as proton pumps. Hence they cannot produce ATP from oxidative phosphorylation as can NADH-Q reductase. These three pathways are:

1. Succinate dehydrogenase in the TCA cycle, part of the succinate-Q reductase complex. Succinate is oxidized to fumarate.
2. Fatty acyl-CoA dehydrogenase in the β-oxidation of fatty acids.
3. Glycerol phosphate dehydrogenase oxidizes glycerol-3-P to DHAP.

NADH and NAD^+ cannot cross the mitochondrial membrane. NADH produced by glycolysis in the cytoplasm is recycled to NAD^+ through two shuttles: the glycerol phosphate shuttle and the malate-aspartate shuttle.

In the glycerol phosphate shuttle, NADH in the cytoplasm reduces DHAP to glycerol-3-P, which enters the mitochondria. The glycerol phosphate dehydrogenase complex in the mitochondria regenerates DHAP, which can return to the cytoplasm. In the process, ubiquinone is reduced to ubiquinol. Transfer of its electrons to cytochrome reductase and then cytochrome oxidase leads to formation of 2 moles of ATP from every NADH.

In the liver and heart the malate-aspartate shuttle can be used. In this shuttle, oxaloacetate is reduced to malate by NADH in the cytoplasm. Malate enters the mitochondria, where it is oxidized by NAD^+, regenerating NADH and oxaloacetate. Oxaloacetate, however, cannot cross the mitochondrial membrane. Instead it is transaminated to aspartate, which can return to the cytoplasm. The malate-asparate shuttle produces 3 moles of ATP per NADH, in contrast to the glycerol phosphate shuttle.

Oxidative Phosphorylation

Oxidative phosphorylation involves the production of ATP from ADP and P_i by harnessing the energy released during electron transport. The flow of electrons through the respiratory chain creates a proton gradient across the inner mitochondrial membrane. This gradient is composed of a pH gradient and a membrane potential; the cytoplasmic side is more acidic and has a positive charge compared to the inner side of the mitochondrial membrane. Protons are pumped out of the inner membrane at each of the three protein complexes transferring electrons from NADH to oxygen: NADH-Q reductase, cytochrome reductase, and cytochrome oxidase. The enzyme ATP synthase phosphorylates ADP to ATP at each of these three protein complexes. The movement of protons back into the mitochondria releases ATP attached to ATP synthase. Without the proton gradient, ATP would stay bound to ATP synthase.

ATP formed within the mitochondria is used primarily in the cytoplasm. ATP and ADP, however, cannot freely cross the mitochondrial membrane on their own. ATP-ADP translocase, a transport protein in the mitochondrial membrane, couples the entry of ADP to the exit of ATP. Transport is driven by the membrane potential generated from electron transport.

Electron transport from NADH to oxygen produces 3 moles of ATP. The net reaction is:

$$NADH \longrightarrow NAD^+ + H^+ + 2e^- \qquad E = -E^{0\prime} = +0.320 \text{ V}$$
$$\tfrac{1}{2}O_2 + 2H^+ + 2e^- \longrightarrow H_2O \qquad E = E^{0\prime} = +0.816 \text{ V}$$

$$\textit{Net:} \; NADH + \tfrac{1}{2}O_2 + H^+ \longrightarrow NAD^+ + H_2O \quad \Delta E = +1.14 \text{ V}$$

$$\Delta G^{0\prime} = -23.1(n)(\Delta E^{0\prime}) = -23.1(2)(1.14) = -52.7 \text{ kcal/mole}$$

The phosphorylation of 3 moles of ADP to ATP requires 21.9 kcal (3 × 7.3 kcal/mole). Hence the efficiency of electron transport is about 40% (21.9/52.7).

The oxidation of glucose to CO_2 and water yields between 36 and 38 moles of ATP, depending on which shuttle is used to transfer the reducing power of cytoplasmic NADH into the mitochondria. Thus the net energy gain from glycolysis to pyruvate is either 6 or 8 moles of ATP (6 with the glycerol phosphate shuttle and 8 with the malate-aspartate shuttle). The pyruvate-dehydrogenase reaction generates 2 moles of NADH in the mitochondria per mole of glucose, which is worth 6 moles of ATP. The TCA cycle receives two acetate groups from every molecule of glucose converted to acetyl-CoA. Two turns of the TCA cycle, therefore, produce 6 moles of NADH (worth 18 moles of ATP), 2 moles of FADH$_2$ (worth 4 moles of ATP), and 2 moles of GTP (worth 2 moles of ATP), a total energy gain equivalent to 24 moles of ATP from oxidizing two acetate groups to CO$_2$ and water. *Therefore, the complete oxidation of glucose to CO$_2$ and water yields between 36 and 38 moles of ATP.* This is compared with the meager 2 moles of ATP produced in anaerobic glycolysis, when the TCA cycle and oxidative phosphorylation cannot operate. Thus, anaerobic muscle must convert glucose to lactate at a terrific pace to compensate for this low energy yield.

Ordinarily electron transport is tightly coupled to oxidative phosphorylation; the oxidation of FADH$_2$ and NADH is linked to the phosphorylation of ADP to ATP. This regulation is termed **respiratory control.** The rate-controlling factor in electron transport is the availability of ADP. Under normal circumstances, O$_2$, NADH, FADH$_2$, and P$_i$ are present in sufficient quantity to allow electron transport; their concentrations do not influence the rate of electron transport. Electron transport is tightly coupled to oxidative phosphorylation; when ADP disappears, electron transport stops.

When electron transport proceeds without concomitant ATP production, the reactions in the mitochondria are said to be uncoupled. Uncoupling follows damage to mitochondria or exposure to uncoupling agents such as 2,4-dinitrophenol (DNP). DNP brings protons across the mitochondrial membrane, thereby neutralizing the proton gradient. The energy released in electron transport by uncoupled mitochondria is liberated as heat.

The P/O ratio for a substrate is defined as the ratio of P$_i$ consumed in ATP formation per oxygen atom consumed. The P/O ratio for mitochondrial NADH is 3, while for FADH$_2$ it is 2. Uncoupled mitochondria have a P/O ratio of 0 for any substrate.

Substances blocking electron transport illustrate the coupling of oxidative phosphorylation to electron transport. Amobarbital and rotenone block NADH-Q reductase. NADH cannot be utilized, but the succinate reductase complex continues to produce 2 moles of ATP per molecule of succinate. Antimycin A blocks cytochrome reductase. Hence neither NADH nor succinate can be utilized to produce ATP. Carbon monoxide and cyanide block cytochrome oxidase, thereby stopping the entire electron transport chain.

Problems
Problem 1
Which coenzyme is not required in the pyruvate and α-ketoglutarate dehydrogenase reactions?

A. Biotin.

B. Lipoic acid.

C. NAD^+.
D. FAD.
E. TPP.

Problem 2
Uncouplers of oxidative phosphorylation like dinitrophenol:

A. Block electron transport.
B. Increase the P/O ratio.
C. Bind to ATP synthase.
✓ D. Reduce the proton gradient across the inner mitochondrial membrane.
E. Inactivate ATP-ADP translocase.

Problem 3
Regarding the pyruvate dehydrogenase (PDH) reaction:

A. PDH kinase removes phosphate from inactive PDH, thereby activating it.
B. PDH phosphatase is stimulated by high ratios of ATP/ADP, NADH/NAD, and acetyl-CoA/CoA.
✓ C. The action of PDH kinase resembles that of the enzyme that activates glycogen phosphorylase.
D. PDH is a flavin-linked dehydrogenase.
E. PDH is located in the cytosol.

Problem 4
Which of the following does not control the rate of the citric acid cycle?

A. Isocitrate dehydrogenase.
✓ B. Fumarase.
C. Citrate synthase.
D. α-Ketoglutarate dehydrogenase.
E. Availability of oxaloacetate.

Problem 5
The glycerol phosphate shuttle, as opposed to the malate-aspartate shuttle:

✓ A. Generates 2 moles of mitochondrial NADH for every cytoplasmic NADH oxidized to NAD^+.
B. Assists ATP-ADP translocase.
C. Leads to the production of 38 moles of ATP from the complete oxidation of glucose.
D. Utilizes lysosomal enzymes.

Problem 6
Which step of the TCA cycle utilizes phosphate?

A. Fumarase.
B. Aconitase.
C. Isocitrate dehydrogenase.

D. Succinate dehydrogenase.
√ E. Succinyl-CoA synthetase.

Problem 7
In healthy nonischemic cells, the rate of electron transport is governed by which factor?

A. NADH.
√ B. ADP.
C. ATP.
D. P_i.
E. Acetyl-CoA.

Answers

1. A. Biotin is used in carboxylation reactions.
2. D. Uncouplers reduce the P/O ratio to 0.
3. C. PDH phosphatase removes phosphate from inactive PDH, thereby activating it. PDH kinase is stimulated by high ratios of ATP/ADP, NADH/NAD, and acetyl-CoA/CoA. PDH is an NAD-linked dehydrogenase. PDH is a mitochondrial enzyme.
4. B.
5. A. Neither shuttle uses lysosomal enzymes.
6. E.
7. B.

References

Devlin, T. M. *Textbook of Biochemistry with Clinical Correlations* (3rd ed.). New York: Wiley-Liss, 1992. Pp. 247–286.

Mathews, C. K., and van Holde, K. E. *Biochemistry*. Redwood City, Calif.: Benjamin/Cummings, 1990. Pp. 467–493, 504–535.

Murray, R. K., Granner, D. K., Mayes, P. A., and Rodwell, V. W. *Harper's Biochemistry* (22nd ed.). Norwalk, Conn.: Appleton & Lange, 1990. Pp. 105–123, 155–162.

Stryer, L. *Biochemistry* (3rd ed.). New York: Freeman, 1988. Pp. 373–426.

Structure and Properties of Lipids and Biomembranes

Lipids are compounds that are soluble in nonpolar solvents such as ether and benzene. Although certain lipids contain ionized groups (e.g., phosphate or choline), the bulk of any lipid molecule is non-polar. Amino acids, proteins, carbohydrates, and nucleotides are too highly polar and ionized to be soluble in these "lipid" solvents.

Structurally, the lipids are quite diverse; there is no common sub-unit in their structure. The primary building blocks in human lipids are fatty acids, glycerol, sphingosine, and sterols.

Fatty Acids

All fatty acids have a single carboxyl group at the end of a hydrocarbon chain, which makes them weak acids. Acetic acid, CH_3COO^-, is the simplest fatty acid. The three-carbon fatty acid (C3) is propionic acid, $CH_3CH_2COO^-$.

Most natural fatty acids have an even number of carbon atoms.

The hydrocarbon chain of fatty acids, represented by $RCOO^-$, can be either saturated (i.e., lacking carbon-carbon double bonds) or un-saturated.

The two most abundant saturated fatty acids in humans are palmitic acid (C16) and stearic acid (C18).

Oleic acid (C18) and palmitoleic acid (C16) compose the bulk of the monounsaturated, or monoenoic, fatty acids in humans. Both have a carbon-carbon double bond between carbons 9 and 10.

Polyunsaturated, or polyenoic, fatty acids include linoleic acid (C18) with two double bonds, linolenic acid (C18) with three, and arachidonic acid (C20) with four double bonds. Linoleic acid and lin-olenic acid are termed essential fatty acids because they cannot be synthesized by mammals. Arachidonic acid is produced in humans from linolenic acid.

Polyunsaturated fatty acids are the precursors for the prostaglandin hormones. All prostaglandins have 20 carbon atoms with a 5-carbon ring. As shown below, part of the straight chain of arachidonic acid is folded into a five-carbon ring while two of its double bonds are oxygenated to generate prostaglandin E_2 (PGE_2). The bonds drawn as dots project downward perpendicular to the page.

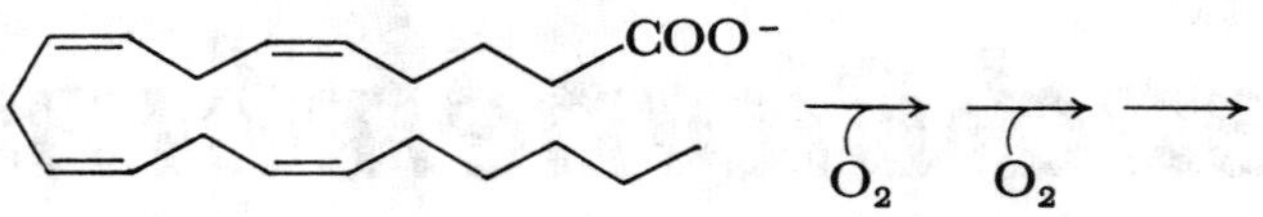

Arachidonic acid

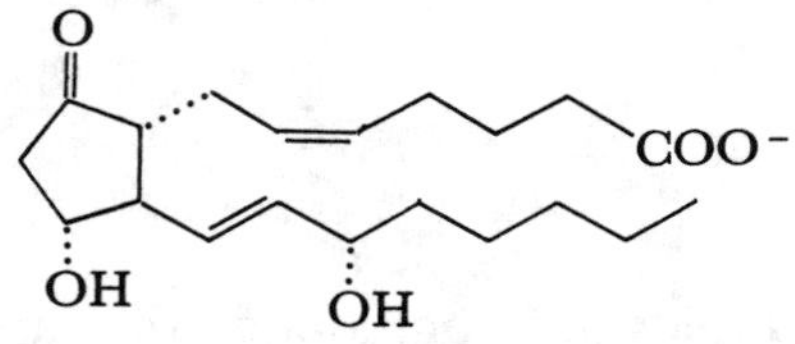

PGE_2

Acylglycerols

The **acylglycerols** are esters of fatty acids bound to the sugar alcohol glycerol. They are also called **neutral fats,** because the carboxyl groups of the fatty acids are bound in ester linkage and can no longer function as acids.

The triacylglycerols, or **triglycerides,** are the principal storage fats in humans. They are named according to their fatty-acid content; tripalmitin contains glycerol and three palmitate chains, while stearodiolein contains glycerol with one stearate and two oleate chains.

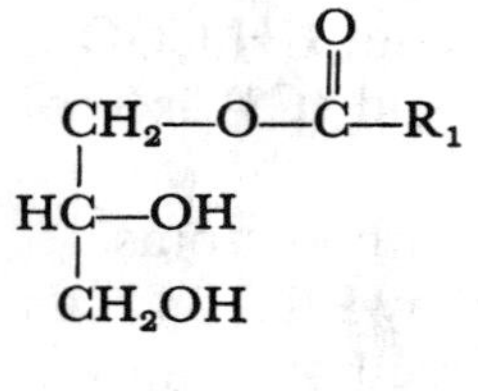

1-Acylglycerol 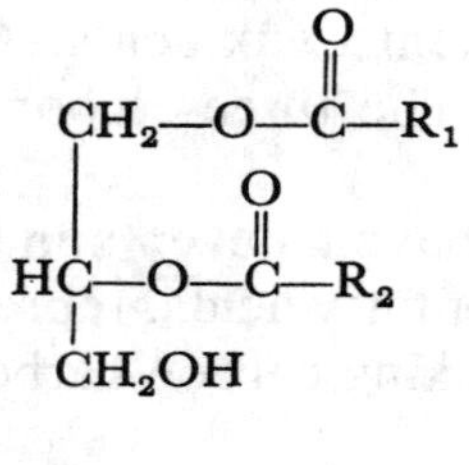 1,2-Diacylglycerol 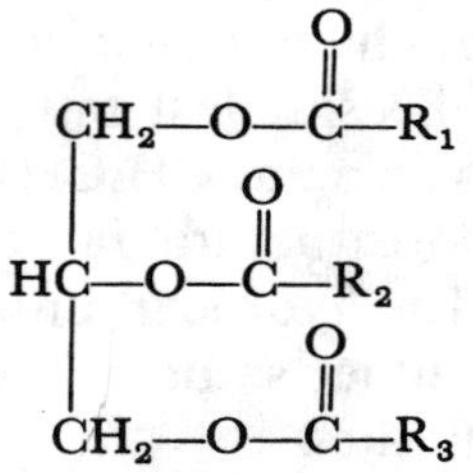 Triacylglycerol

In humans, triglycerides are hydrolyzed to glycerol and free fatty acids (FFA) by lipase enzymes.

In industry, neutral fats are hydrolyzed with NaOH, or **saponified,** to create water-soluble fatty-acid soaps ($RCOO^-Na^+$) and glycerol. Soaps have a detergent action because after the nonpolar R groups in the soaps bind to lipids in the skin or clothing, the ionized carboxyl group can pull this bound lipid into the water phase.

Calcium binds to free fatty acids in the intestine to generate insoluble calcium–fatty-acid soaps, thereby blocking their absorption. In acute pancreatitis, lipase released from the pancreas into the bloodstream hydrolyzes triacylglycerols and creates calcium–fatty-acid soaps.

Phosphoglycerides

The **phosphoglycerides** are phosphate esters of diglycerides. Glycerol-3-phosphate is the structural backbone of the phosphoglycerides. Two fatty acids are esterified to glycerol-3-phosphate to produce the **phosphatidic acids,** which are intermediates in the synthesis of triacylglycerols and various other phosphoglycerides.

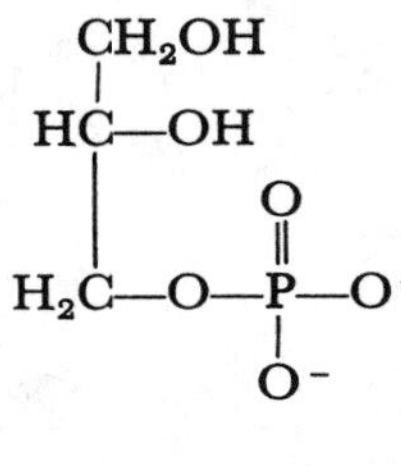

$$
\begin{array}{ll}
CH_2OH & \\
| & \\
HC-OH & \\
| & \\
H_2C-O-P-O^- & \\
\end{array}
\qquad
\begin{array}{l}
H_2C-O-C(=O)-R_1 \\
HC-O-C(=O)-R_2 \\
H_2C-O-P(=O)(O^-)-O^-
\end{array}
$$

Glycerol-3-phosphate 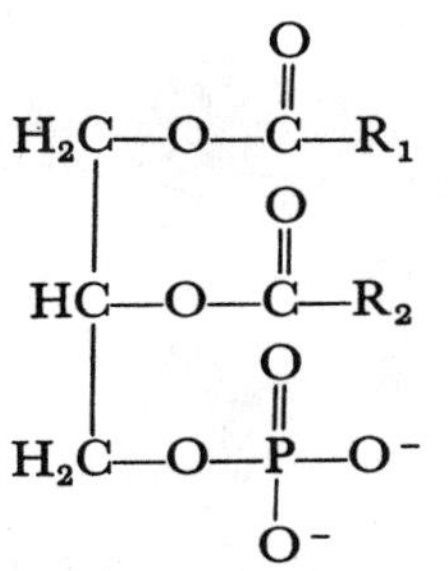   Phosphatidic acid

By esterifying choline, or trimethylethanolamine $HOCH_2CH_2$ $^+N(CH_3)_3$, to the phosphoric acid portion of phosphatidic acid, one gets phosphatidylcholine. Also called lecithin, the phosphatidylcholines play an essential role in reducing surface tension in lung alveoli; they are surfactants or surface-acting agents. Respiratory distress syndrome (RDS) of the newborn, which is common in premature infants, results from a lack of this surfactant in the lung. The lung is stiff, expands with difficulty, and has many collapsed portions. To predict the likelihood of RDS in high-risk pregnancies, obstetricians commonly perform amniocentesis for laboratory determination of the ratio of phosphatidylcholine (lecithin) to sphingomyelin in the amniotic fluid (L/S ratio). The higher the L/S ratio, the more surfactant is present to allow the lung to expand normally.

Other important human phosphoglycerides are phosphatidylserine and phosphatidylethanolamine. The common serologic test for syphilis, the Venereal Disease Research Laboratory (VDRL) test, utilizes cardiolipin, a diphosphatidyl glycerol, as the antigen.

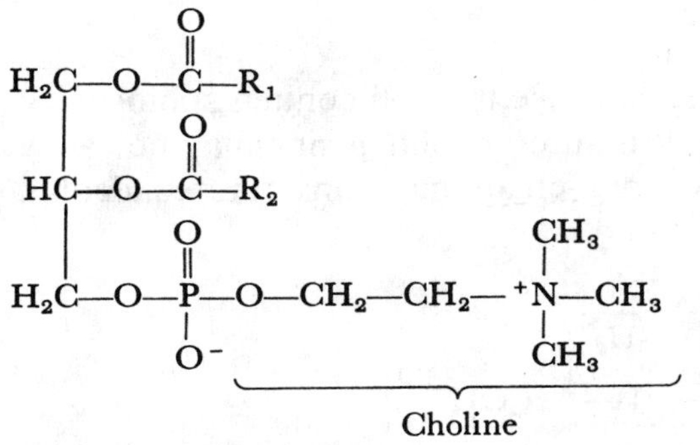

$$
\begin{array}{l}
H_2C-O-C(=O)-R_1 \\
HC-O-C(=O)-R_2 \\
H_2C-O-P(=O)(O^-)-O-CH_2-CH_2-\,^+N(CH_3)_3
\end{array}
$$

Choline

Phosphatidylcholine

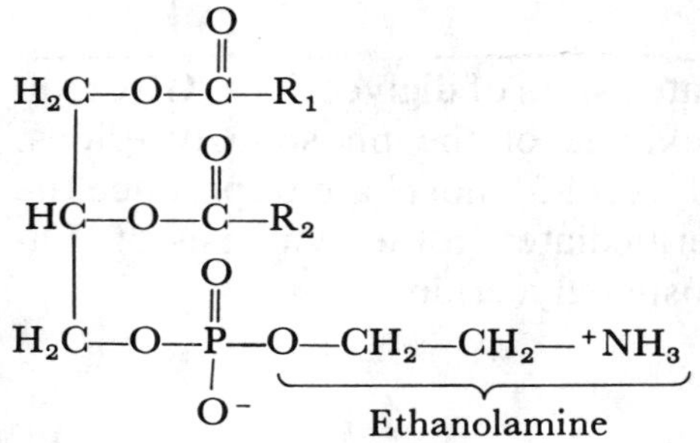

$$
\begin{array}{l}
H_2C-O-\overset{\displaystyle O}{\overset{\|}{C}}-R_1 \\[4pt]
HC-O-\overset{\displaystyle O}{\overset{\|}{C}}-R_2 \\[4pt]
H_2C-O-\overset{\displaystyle O}{\underset{\displaystyle O^-}{P}}-O-CH_2-CH_2-{}^+NH_3
\end{array}
$$

$\underbrace{\qquad\qquad}_{\text{Ethanolamine}}$

Phosphatidylethanolamine

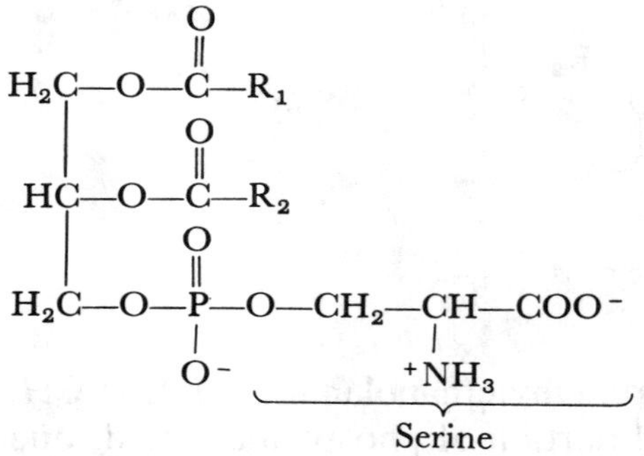

$$
\begin{array}{l}
H_2C-O-\overset{\displaystyle O}{\overset{\|}{C}}-R_1 \\[4pt]
HC-O-\overset{\displaystyle O}{\overset{\|}{C}}-R_2 \\[4pt]
H_2C-O-\overset{\displaystyle O}{\underset{\displaystyle O^-}{P}}-O-CH_2-\underset{\displaystyle {}^+NH_3}{CH}-COO^-
\end{array}
$$

Serine

Phosphatidylserine

The highly polar phosphate, choline, and serine groups of the phosphoglycerides make these compounds water-soluble, while their fatty acyl groups confer solubility in nonpolar agents. Hence, they can serve to cement lipids in membranes and lipoproteins to the polar proteins and carbohydrates.

Phosphatidylcholine is the major storage form for choline inside the brain. Choline acetyltransferase synthesizes the cholinergic neurotransmitter, acetylcholine:

$$\text{Choline} + \text{acetyl-CoA} \xrightarrow{\text{Choline acetyltransferase}} \text{acetylcholine} + \text{CoA-SH}$$

Sphingolipids

The **sphingolipids** are so named because they all contain sphingosine or one of its derivatives. The structure of sphingosine may be easily identified, because it contains a long-chain, monounsaturated alcohol bound to ethanolamine:

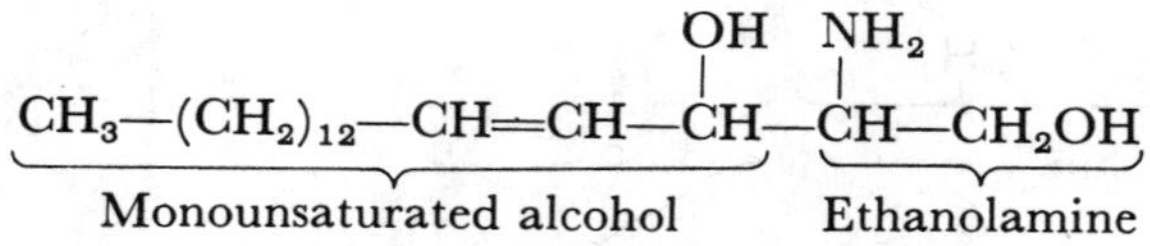

$$
\underbrace{CH_3-(CH_2)_{12}-CH{=}CH-\underset{\displaystyle OH}{CH}}_{\text{Monounsaturated alcohol}}-\underbrace{\underset{\displaystyle NH_2}{CH}-CH_2OH}_{\text{Ethanolamine}}
$$

Sphingosine

In addition to sphingosine, *all sphingolipids contain a fatty acid. None contains glycerol.*

Sphingolipids abound in the nervous system as components of myelin and other structural lipids. They occur to a lesser extent in the liver, spleen, and bone marrow.

Ceramides, the simplest sphingolipids, consist of a fatty acid bound to sphingosine. In humans, ceramides function principally as intermediates in the synthesis of other sphingolipids; all other sphingolipids contain ceramide.

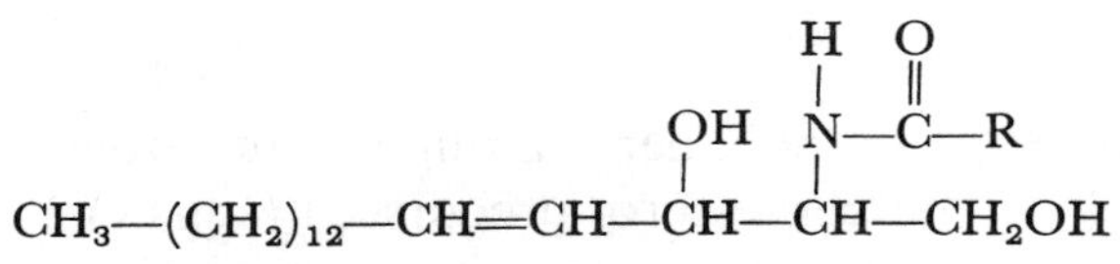

$$CH_3-(CH_2)_{12}-CH{=}CH-\underset{\underset{\displaystyle OH}{|}}{CH}-\underset{\underset{\displaystyle \overset{H}{\underset{|}{N}}-\overset{O}{\overset{\|}{C}}-R}{|}}{CH}-CH_2OH$$

Ceramide

By joining choline phosphate or ethanolamine phosphate to ceramides, one generates the **sphingomyelins,** which are important components of the myelin sheath surrounding the fastest conducting nerve fibers.

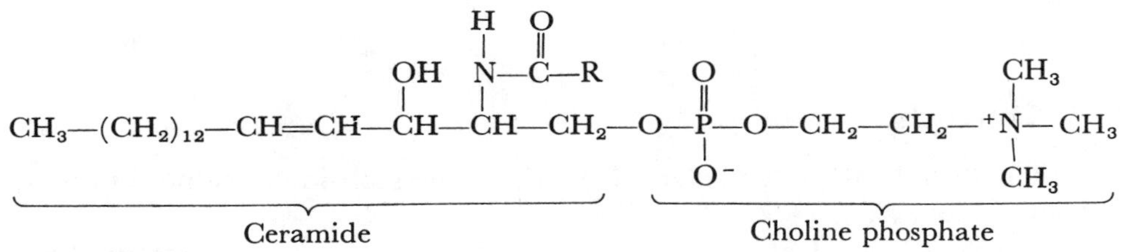

$$CH_3-(CH_2)_{12}-CH{=}CH-CH-CH-CH_2-O-P-O-CH_2-CH_2-{}^+N-CH_3$$

Ceramide   Choline phosphate

Sphingomyelin

Cerebrosides consist of a hexose sugar, such as glucose or galactose, bound to a ceramide. These ceramide-monosaccharides are also part of the myelin sheath.

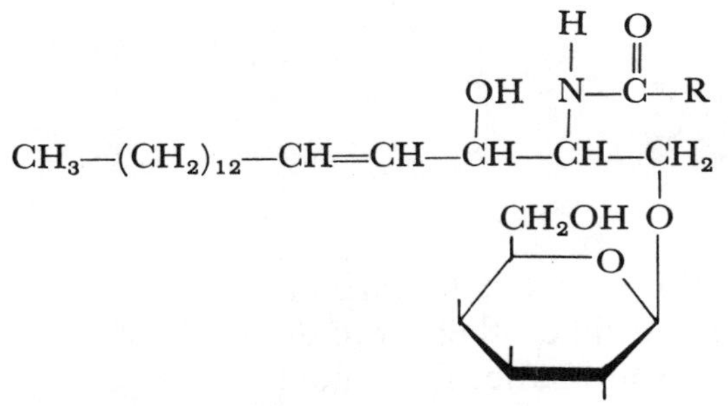

$$CH_3-(CH_2)_{12}-CH{=}CH-CH-CH-CH_2$$

Galactocerebroside

Gangliosides consist of ceramide bound to an oligosaccharide that contains an acidic sugar such as *N*-acetylneuraminic acid.

Sulfatides are sulfated cerebrosides, or cerebroside-sulfate esters.

Deficiencies of the enzymes involved in sphingolipid catabolism lead to the pathologic accumulation of sphingolipids. Chapter 11 on lipid metabolism gives further details on the accumulation of gluco-cerebrosides in Gaucher's disease, ceramide-trisaccharides in Fabry's disease, sphingomyelin in Niemann-Pick disease, sulfatides in metachromatic leukodystrophy, and GM_2 gangliosides in Tay-Sachs disease.

Terpenes

The **terpenes** are a class of isoprene polymers. Examples of terpene derivatives in humans include the cholesterol precursors (squalene, geraniol, and farnesol), provitamin A (β-carotene), and vitamin A_1 (retinol).

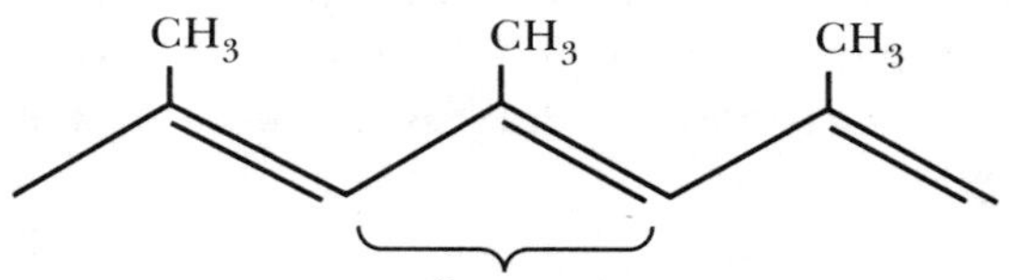

Isoprene
subunit

Steroids

The broad category of **steroids** includes the steroid hormones, sterols, and bile acids.

The essential structural nucleus of the steroids consists of three fused cyclohexane rings (A–C) joined to a cyclopentane ring (D). Except for the estrogens, steroids do not contain aromatic rings. Carbons 3 and 17 always have side groups.

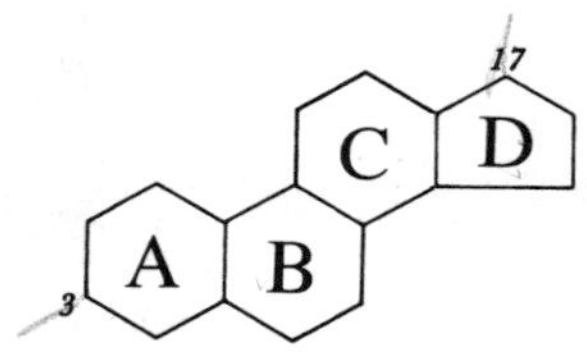

Steroid nucleus

Estrogens, the ovarian steroids, contain 18 carbon atoms (carbon 18 is found in a methyl group). Unlike other steroids, the A ring of the steroid nucleus of estrogen is aromatic. Estradiol (next page) has OH groups attached to carbons 3 and 17.

Androgens, produced in the adrenal cortex and the testes, have 19 carbon atoms (carbons 18 and 19 are in methyl groups). Testosterone is one of the more potent androgens. Because of its keto group at carbon 3, testosterone has a double bond in its initial cyclohexane ring. Dehydroepiandrosterone (DHA), unlike testosterone, is a 17-ketosteroid.

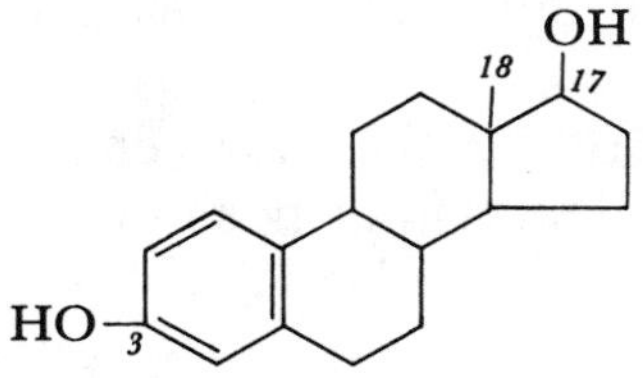

Estradiol

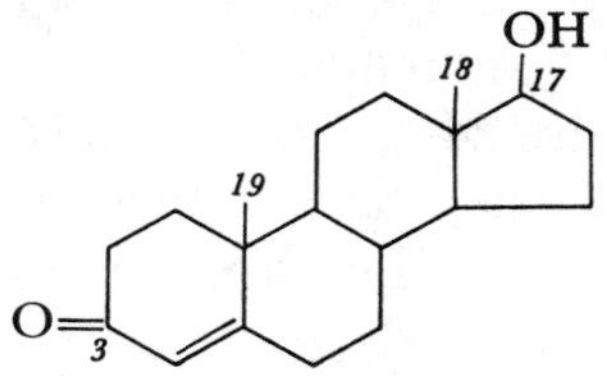

Testosterone

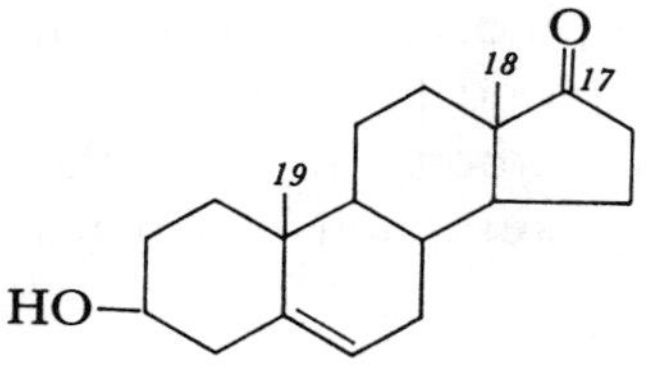

DHA

Progesterone, synthesized in the corpus luteum, has 21 carbons, as do the adrenocortical steroids such as corticosterone and cortisol. Progesterone has an acetyl group joined at carbon 17.

The adrenal cortex produces **glucocorticoids,** which raise the serum glucose level, and **mineralocorticoids,** which promote renal sodium retention. Cortisol is a potent glucocorticoid with weak mineralocorticoid activity, whereas aldosterone is a potent mineralocorticoid (presumably due to its aldehyde group at carbon 18) but a weak glucocorticoid. A hydroxyl or keto group at carbon 11 is found to correlate with glucocorticoid activity. The urinary 17-hydroxycorticosteroid assay detects all 21-carbon steroids with a 17-OH group, such as cortisol.

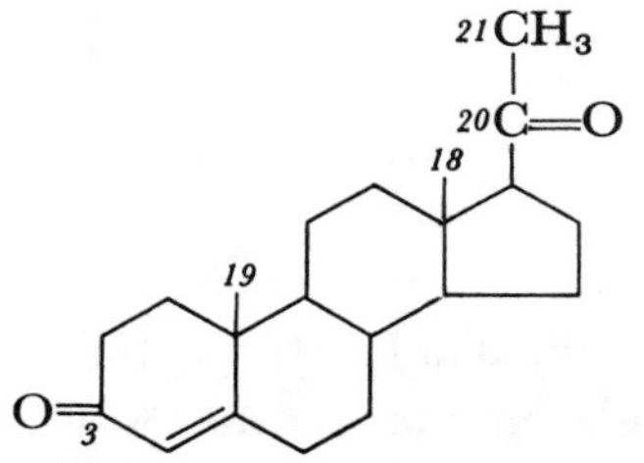

Progesterone

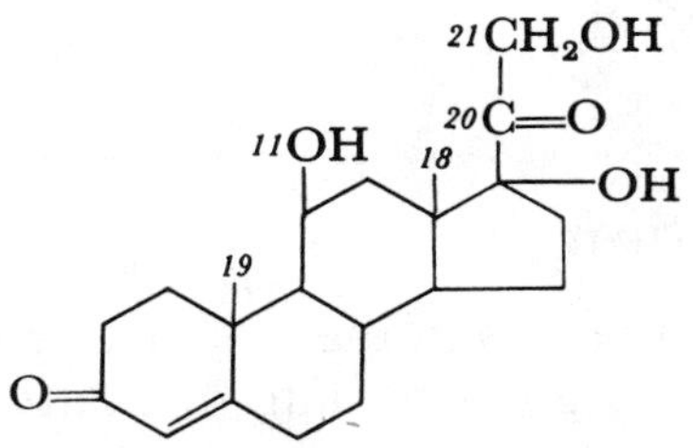

Cortisol

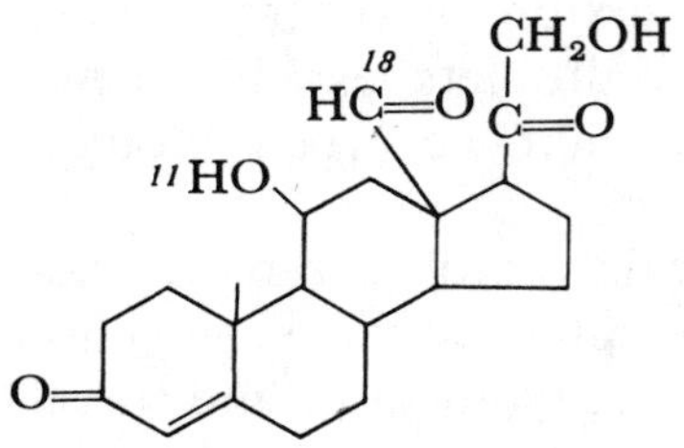

Aldosterone

Deficiency of the 21-hydroxylase enzyme in congenital adrenal hyperplasia leads to underproduction of cortisol and overproduction of 19-carbon androgens, with a consequent rise in urinary 17-ketosteroid excretion. Excess androgens cause virilization of females and precocious puberty in males.

The **bile acids** are 24-carbon steroids secreted into the bile to emulsify dietary fats. They have a five-carbon side chain at position 17 that contains a carboxyl group, making them acidic. Cholic acid is a major human bile acid.

The **sterols** are steroids with 27 to 29 carbon atoms and an OH group at carbon 3. Cholesterol, the major human sterol, is the precursor to all the steroid hormones. In addition, it occurs in high concentration in the brain. Most cholesterol in blood is bound to unsaturated fatty acid through the OH group at carbon 3 to form cholesterol esters. The vitamin D precursors, ergosterol and 7-dehydrocholesterol, are also sterols.

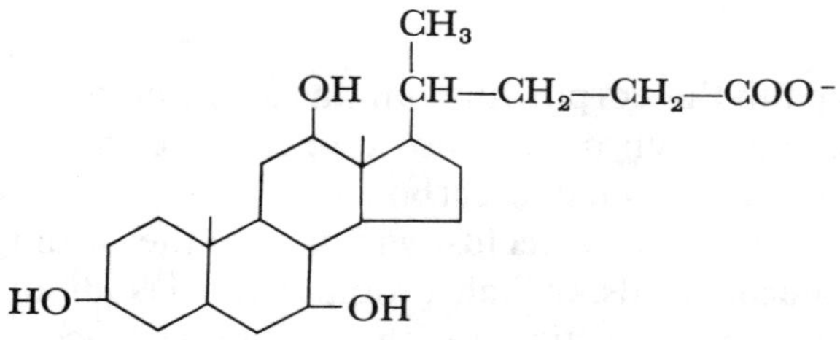

Cholic acid

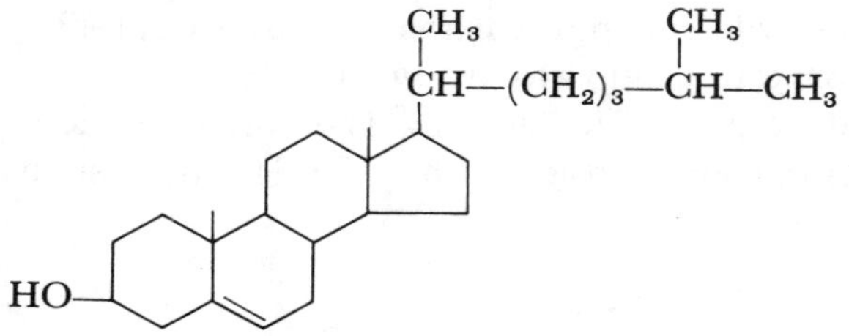

Cholesterol

Lipid Digestion

Lipid digestion begins in the stomach, where gastric lipase partially hydrolyzes triacylglycerols into free fatty acids, glycerol, and mono- and diacylglycerols. In the duodenum, pancreatic lipase continues this process. Phospholipases remove fatty acids from phosphoglycerides. Bile salts solubilize or emulsify this mixture of free fatty acids, monoacylglycerols, and diacylglycerols into droplets less than 1 micron in diameter termed **micelles,** which are readily absorbed throughout the small intestine.

Inside the intestinal mucosal cells, triacylglycerols are re-synthesized and combined with β-lipoproteins, phosphoglycerides, and cholesterol to form **chylomicrons,** which enter the lymphatics and travel through the thoracic duct to reach the bloodstream.

Unlike long-chain fatty acids, the medium- and short-chain fatty acids (i.e., those with less than 12 carbons) can be well absorbed without bile salts. They enter the portal venous system directly and travel to the liver, instead of traveling through the lymphatics as triacylglycerols in chylomicrons. Medium- and short-chain fatty-acid preparations are used as a source of fatty acids for patients with lipid malabsorption disorders, such as cystic fibrosis.

Bile salts also emulsify cholesterol to hasten its absorption. Once absorbed, cholesterol is esterified to unsaturated fatty acids to create cholesterol esters.

Cholestyramine is a resin used clinically to lower the serum cholesterol level. It binds to bile salts, thereby blocking cholesterol absorption and augmenting the hepatic transformation of cholesterol to bile salts.

Lipoproteins

Lipids must bind to proteins to make them water-soluble for transport in the blood. The protein components of lipoproteins are termed **apolipoproteins.** Free fatty acids avidly bind to serum albumin and will displace albumin-bound drugs from their binding sites.

Two laboratory techniques are used to separate lipoproteins from one another: ultracentrifugation separates them according to their differing densities and electrophoresis separates them on the basis of their varying net charges.

Chylomicrons are the least dense lipoproteins because they consist mainly of triglycerides with small amounts of cholesterol, phospholipids, and proteins. They do not migrate when subjected to electrophoresis, because of their high triglyceride content (triacylglycerols have no charge). After a fatty meal, the blood appears milky due to the high concentration of chylomicrons. Lipoprotein lipase hydrolyzes triglycerides bound in lipoproteins such as chylomicrons and VLDL, yielding monoglycerides and free fatty acids. Heparin, an anticoagulant, also helps to clear chylomicrons from the blood, perhaps by stimulating lipoprotein lipase. The inherited absence of lipoprotein lipase causes hyperchylomicronemia, termed Fredrickson's type-I hyperlipoproteinemia.

Very-low-density lipoproteins (VLDL) also contain principally triglycerides, but they have a greater protein, phospholipid, and cholesterol content than chylomicrons. Their protein and phospholipid content makes them charged so that they migrate just before the β-globulins in electrophoresis; hence, they are termed **pre-β-lipoproteins.** VLDL are synthesized in the liver. This VLDL fraction is markedly elevated in type-IV hyperlipoproteinemia.

Low-density lipoproteins (LDL) contain mainly cholesterol, in contrast to the content of chylomicrons and VLDL, which are mainly triglyceride. LDL also contain appreciable amounts of proteins (chiefly apolipoprotein B-100), phospholipids, and triglycerides. Because they migrate with the β-globulins, they are termed β-lipoproteins. The LDL fraction is markedly elevated in type-II hyperlipoproteinemia.

Intermediate-density lipoproteins (IDL) contain less cholesterol and more protein than LDL. The degradation of triacylglycerol in VLDL by lipoprotein lipase produces IDL and HDL. IDL are in turn degraded to LDL.

High-density lipoproteins (HDL) contain mainly protein (apolipoproteins A1 and A2) and phospholipid. They contain significant amounts of cholesterol, but they have little triglyceride. The high-protein, low-triglyceride content makes them very dense. They are termed **α-lipoproteins** and are separated from other lipoproteins by electrophoresis.

Biological Membranes

Biological membranes are thin sheetlike structures composed mainly of lipid and protein. Membrane lipids create the permeability barrier, while membrane proteins serve as pumps, enzymes, receptors, and energy transducers. Membranes create compartments ranging from mitochondria and nuclei to entire cells.

Three main classes of lipids are found in biological membranes: phospholipids, glycolipids, and cholesterol. The phospholipids are based on either glycerol or sphingosine. Glycerol-based phospholipids include phosphatidylcholine, phosphatidylinositol, and phosphatidylethanolamine. Sphingomyelin is one of the sphingosine-based phospholipids. Membrane glycolipids include cerebrosides and gangliosides.

Membrane lipids spontaneously form bilayers. Bilayers are like sandwiches, in that they have two layers, each one-molecule thick. The hydrophobic components face inward, whereas the hydrophilic components face outward. In this drawing, circles depict the polar or hydrophobic components.

Thus on the outer surfaces of lipid bilayers one finds the OH groups of cholesterol, sugar residues from glycolipids, and phosphorylated compounds such as phosphoryl choline. In the center of the bilayers one finds hydrophobic compounds such as cholesterol (except for its OH group) and the hydrocarbon chains of fatty acids and sphingosine.

Membrane lipids have two principal functions: (1) as a permeability barrier and (2) as a solvent for membrane proteins. Proteins, linked noncovalently to membrane lipids, assume the other membrane functions.

Membranes are asymmetric in both structure and function. Glycoproteins acting as hormone receptors, for example, are located on the outside of cell membranes. The Na^+-K^+ pump transports Na^+ out of cells in exchange for K^+.

Signal Transduction

The transmission of chemical messages from hormones into cells is known as **signal transduction.** There are four methods of signal transduction:

1. Production of cyclic AMP by adenyl cyclase.
2. Production of inositol triphosphate and diacylglycerols from phosphatidylinositol biphosphate.
3. Stimulation of tyrosine kinase in receptors.
4. Transport to the cell nucleus.

Epinephrine and glucagon stimulate guanyl-nucleotide–binding proteins (also known as G proteins) in hormone receptors, which in turn stimulate adenyl cyclase to convert ATP to cyclic AMP. G proteins exist in two forms: an inactive GDP form and an active GTP form. Hormone binding to receptors converts G proteins to their active GTP forms. The α-subunit of G proteins contains a GTPase that hydrolyzes the active form back to the inactive form, thereby terminating stimulation of adenyl cyclase.

Cyclic AMP activates protein kinase A, which phosphorylates serine and threonine in intracellular proteins relevant to the hormone's message.

The second mechanism for signal transduction involves inositol triphosphate and diacylglycerols. Hormones involved in this mechanism, such as TRH, TSH, and epinephrine, stimulate G proteins in hormone receptors, which activate phospholipase C. Phospholipase C, in turn, cleaves phosphatidylinositol biphosphate to inositol-1,4,5-triphosphate (IP_3) and diacylglycerols. IP_3 then enters the cytosol where it binds to receptors on the endoplasmic reticulum, causing calcium release. Calcium binds to calmodulin and other calcium-binding proteins, triggering the relevant intracellular consequences for the hormone. Diacylglycerols, also second messengers, stimulate protein kinase C, which phosphorylates threonine and serine residues in selected proteins.

The third mechanism for signal transduction involves tyrosine kinase in hormone receptors. Insulin and other growth-promoting hormones bind to receptors containing tyrosine kinase. This binding leads tyrosine kinase to phosphorylate tyrosine residues within itself. The resultant activation of tyrosine kinase leads to phosphorylation of tyrosine residues in cytosolic proteins, thereby enacting the hormonal message.

The final mechanism for signal transduction involves the binding of steroid and thyroid hormones to membrane receptors. The hormone is then transported to the nucleus, where it enhances transcription of specific proteins. Because of the long delays involved in protein synthesis, the effects of steroid and thyroid hormones take hours to develop as opposed to seconds or minutes for the other hormones acting by activating enzymes rather than synthesizing new proteins.

Problems

Problem 1

Steroid hormones:

> A. Act quickly via G proteins.
> B. Stimulate tyrosine kinase in membrane receptors.
> C. Of all types are underproduced in 21-hydroxylase deficiency.
> D. Influence transcription.
> E. Act via a different mechanism than thyroid hormones.

Problem 2

Inositol-1,4,5-triphosphate:

> A. Stimulates phosphorylase C.
> B. Is produced by adenyl cyclase.
> C. Is derived from inosine triphosphate in one enzymatic step.
> D. Inhibits calmodulin.
> E. Acts in concert with diacylglycerols.

Problems 3–8

Match the substances listed below to the structures in Problems 3–8:

> A. Sphingomyelin
> B. Ceramide
> C. Monoacylglycerol
> D. Prostaglandin
> E. Glucocerebroside
> F. Phosphatidylcholine
> G. Triacylglycerol

3.

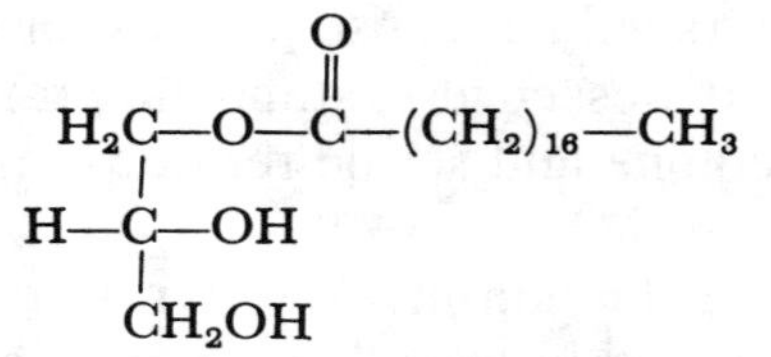

$$H_2C-O-\overset{\overset{\displaystyle O}{\|}}{C}-(CH_2)_{16}-CH_3$$
$$H-\underset{|}{C}-OH$$
$$CH_2OH$$

4.

$$CH_3-(CH_2)_{12}-CH=CH-\underset{\underset{\displaystyle OH}{|}}{CH}-\underset{\underset{\displaystyle N-\overset{\overset{\displaystyle O}{\|}}{C}-(CH_2)_{14}-CH_3}{|}}{\overset{\overset{\displaystyle H}{|}}{CH}}-CH_2OH$$

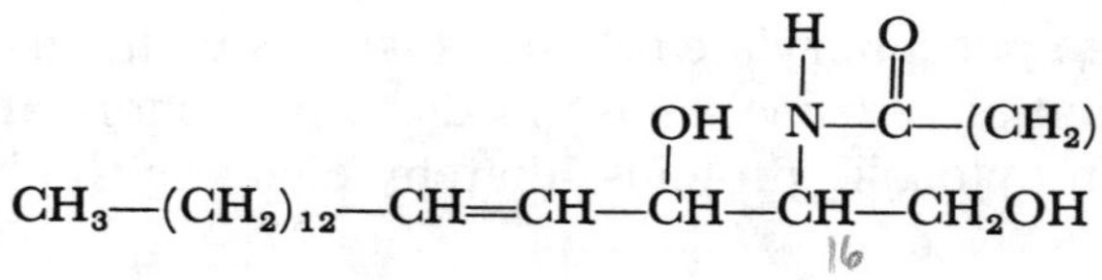

5.

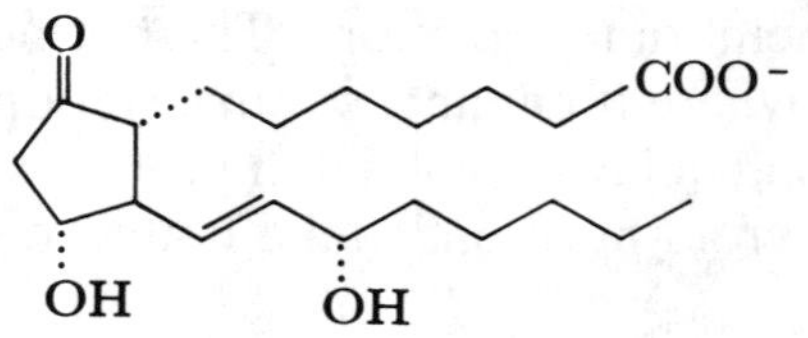

6.

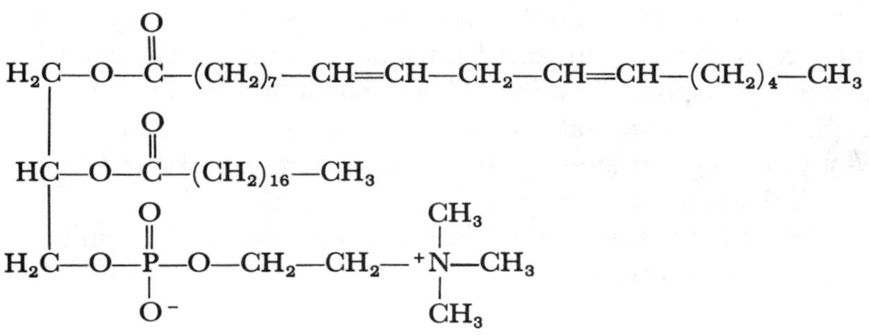

$$H_2C-O-\overset{\overset{\displaystyle O}{\|}}{C}-(CH_2)_7-CH{=}CH-CH_2-CH{=}CH-(CH_2)_4-CH_3$$

$$HC-O-\overset{\overset{\displaystyle O}{\|}}{C}-(CH_2)_{16}-CH_3$$

$$H_2C-O-\overset{\overset{\displaystyle O}{\|}}{\underset{\underset{\displaystyle O^-}{|}}{P}}-O-CH_2-CH_2-{}^+\overset{\overset{\displaystyle CH_3}{|}}{\underset{\underset{\displaystyle CH_3}{|}}{N}}-CH_3$$

7.

$$CH_3-(CH_2)_{12}-CH{=}CH-\overset{\overset{\displaystyle OH}{|}}{CH}-\overset{\overset{\displaystyle \overset{\displaystyle H}{|}\ \overset{\overset{\displaystyle O}{\|}}{N-C}-(CH_2)_{14}-CH_3}{|}}{\underset{\underset{\displaystyle CH_2OH}{}}{CH}}-CH_2$$

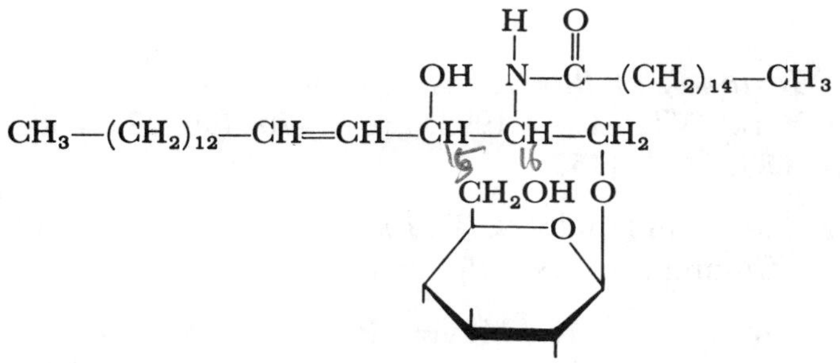

8.

$$CH_3-(CH_2)_{12}-CH{=}CH-\overset{\overset{\displaystyle OH}{|}}{CH}-\overset{\overset{\displaystyle \overset{\displaystyle H}{|}\ \overset{\overset{\displaystyle O}{\|}}{N-C}-(CH_2)_{14}-CH_3}{|}}{CH}-CH_2-O-\overset{\overset{\displaystyle O}{\|}}{\underset{\underset{\displaystyle O^-}{|}}{P}}-O-CH_2-CH_2-{}^+\overset{\overset{\displaystyle CH_3}{|}}{\underset{\underset{\displaystyle CH_3}{|}}{N}}-CH_3$$

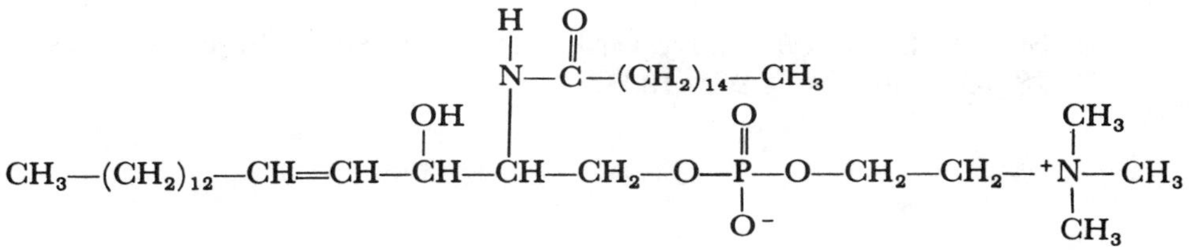

Answers

1. D. Hormones such as glucagon, epinephrine, and TRH act via G proteins. Insulin and other growth hormones stimulate tyrosine kinase in membrane receptors. Cortisol is underproduced in 21-hydroxylase deficiency but other steroid hormones such as adrenal androgens are overproduced. Steroid and thyroid hormones act via the same mechanism; they promote transcription of selective proteins.
2. E. Diacylglycerols, not IP_3, stimulate phosphorylase C. IP_3 is produced by phospholipase C from phosphatidylinositol biphosphate using G proteins as a transducer. IP_3 leads to calcium release, which stimulates calmodulin.
3. C.
4. B. Sphingosine joined to a 16-carbon, saturated fatty acid (palmitic acid) by N-acyl linkage. Sphingosine can be recognized by identifying its monounsaturated alcohol bound to ethanolamine. Ceramides contain sphingosine and a fatty acid.
5. D. Prostaglandin E_2 is an oxygenated derivative of a 20-carbon, polyenoic fatty acid with a five-membered ring and two aliphatic chains.

6. F. Eighteen-carbon, doubly unsaturated fatty acid (linoleic acid) plus an 18-carbon, saturated fatty acid (stearic acid) esterified to glycerol phosphorylcholine. This is phosphatidylcholine.
7. E. Sixteen-carbon, saturated fatty acid (palmitic acid) joined by *N*-acyl linkage to sphingosine, which is in turn bound by glycosidic linkage to glucose. This is a glucocerebroside.
8. A. Palmitic acid and phosphorylcholine bound to sphingosine. This is sphingomyelin.

References

Devlin, T. M. *Textbook of Biochemistry with Clinical Correlations* (3rd ed.). New York: Wiley-Liss, 1992. Pp. 67–70, 196–234, 868–875, 889–894, 1081–1089, 1154–1157.

Mathews, C. K., and van Holde, K. E. *Biochemistry*. Redwood City, Calif.: Benjamin/Cummings, 1990. Pp. 298–331.

Murray, R. K., Granner, D. K., Mayes, P. A., and Rodwell, V. W. *Harper's Biochemistry* (22nd ed.). Norwalk, Conn.: Appleton & Lange, 1990. Pp. 134–145, 441–458.

Stryer, L. *Biochemistry* (3rd ed.). New York: Freeman, 1988. Pp. 283–312, 469–471, 975–1003.

Lipid Catabolism and Biosynthesis

Atherosclerosis, the deposition of lipid plaques on the lining of arteries, is the leading cause of death in America. Because of the association between atherosclerosis and hyperlipoproteinemia (elevated serum lipoproteins), an extensive campaign of medical research has been launched to explore lipid metabolism.

This chapter deals with the metabolism and biosynthesis of six general types of lipids: triglycerides, fatty acids, ketone bodies, cholesterol, phosphoglycerides, and sphingolipids.

Lipolysis

Lipolysis, defined as triglyceride hydrolysis, liberates fatty acids from their main storage depots in the triglycerides. Lipolysis begins with the intestinal hydrolysis of dietary triglycerides by pancreatic lipase. Once absorbed into the intestinal mucosa, the resultant free fatty acids (FFA), glycerol, and monoglycerides are resynthesized into triglycerides, which combine with lesser amounts of protein, phospholipid, and cholesterol to create chylomicrons. Plasma lipoprotein lipase hydrolyzes triglycerides in the chylomicrons into FFA and glycerol. Adipose tissue contains a hormone-sensitive lipase that hydrolyzes its triglycerides.

The glycerol released in lipolysis travels to the liver, where it is phosphorylated to glycerol-3-P. For every 2 moles of glycerol-3-P, 1 mole of glucose can be synthesized in gluconeogenesis. Glycerol-3-P also serves as a precursor to triglyceride synthesis.

Triglyceride Synthesis

The intestinal mucosa differs from other sites of triglyceride synthesis in having a large supply of monoglycerides; each mole of monoglyceride may be combined with 2 moles of fatty-acyl-CoA to produce the corresponding triglyceride. Acyl-CoA synthetase on the outer mitochondrial membrane activates fatty acids by linking them to the SH group of CoA-SH to create fatty-acyl-CoA:

$$R-COO^- + CoA\text{-}SH + ATP \rightleftharpoons R-\overset{\overset{\displaystyle O}{\|}}{C}-S-CoA + AMP + PP_i$$

Pyrophosphate from this reaction is quickly hydrolyzed by pyrophosphatase to 2 P_i, making the net reaction irreversible. Fatty-acid activation costs 2 moles ATP because the endproducts of ATP are AMP and 2 P_i.

In the liver and adipose tissue, triglycerides are synthesized from phosphatidic acids. Two moles of fatty-acyl-CoA add to glycerol-3-P to yield the phosphatidic acid. The second route links 2 moles of

Fig. 11-1 Triglyceride synthesis in liver and adipose tissue.

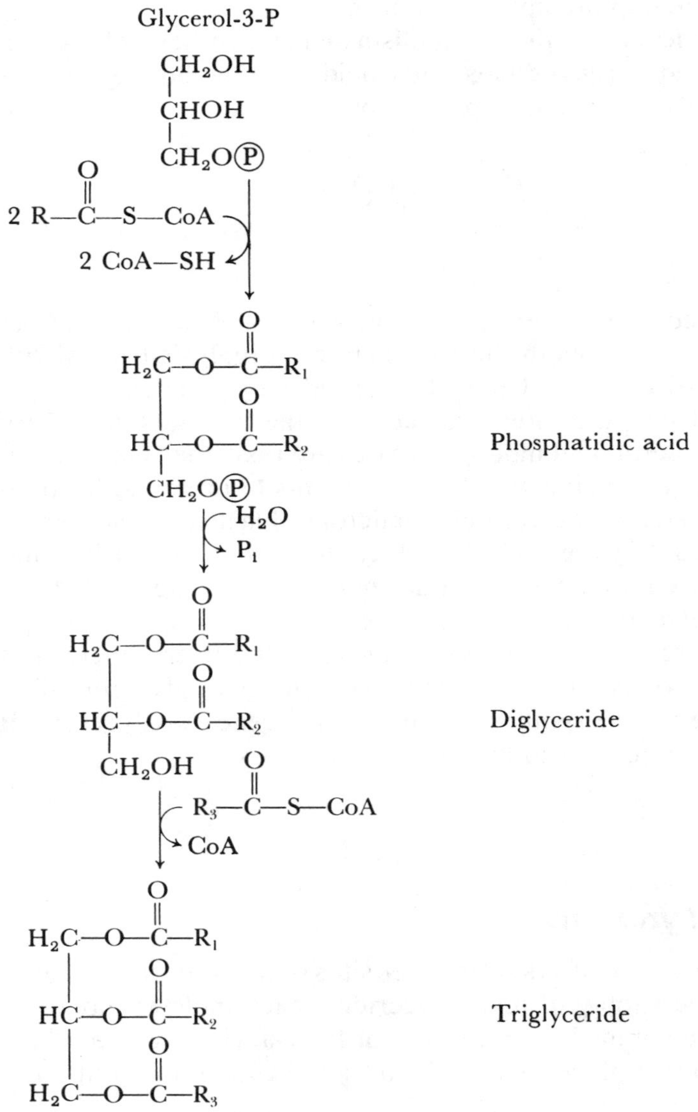

fatty-acyl-CoA to DHAP by means of reducing its keto group with NADPH, thus forming phosphatidic acid, as shown in Figure 11-1. After synthesis via either pathway, phosphate is then cleaved from the phosphatidic acid, leaving a diglyceride. A third fatty-acyl-CoA is esterified to this diglyceride to yield a triglyceride.

Beta-Oxidation of Fatty Acids

As the principal route for catabolizing fatty acids, β-oxidation occurs in the mitochondria, the intracellular "powerhouses." **β-Oxidation** is so named because it oxidizes the β-carbon atom of a fatty acid to a β-keto acid.

α-Oxidation of fatty acids occurs in the human brain. The rare inherited absence of an enzyme required for α-oxidation causes Refsum's disease.

Fig. 11-2 β-Oxidation of fatty acids.

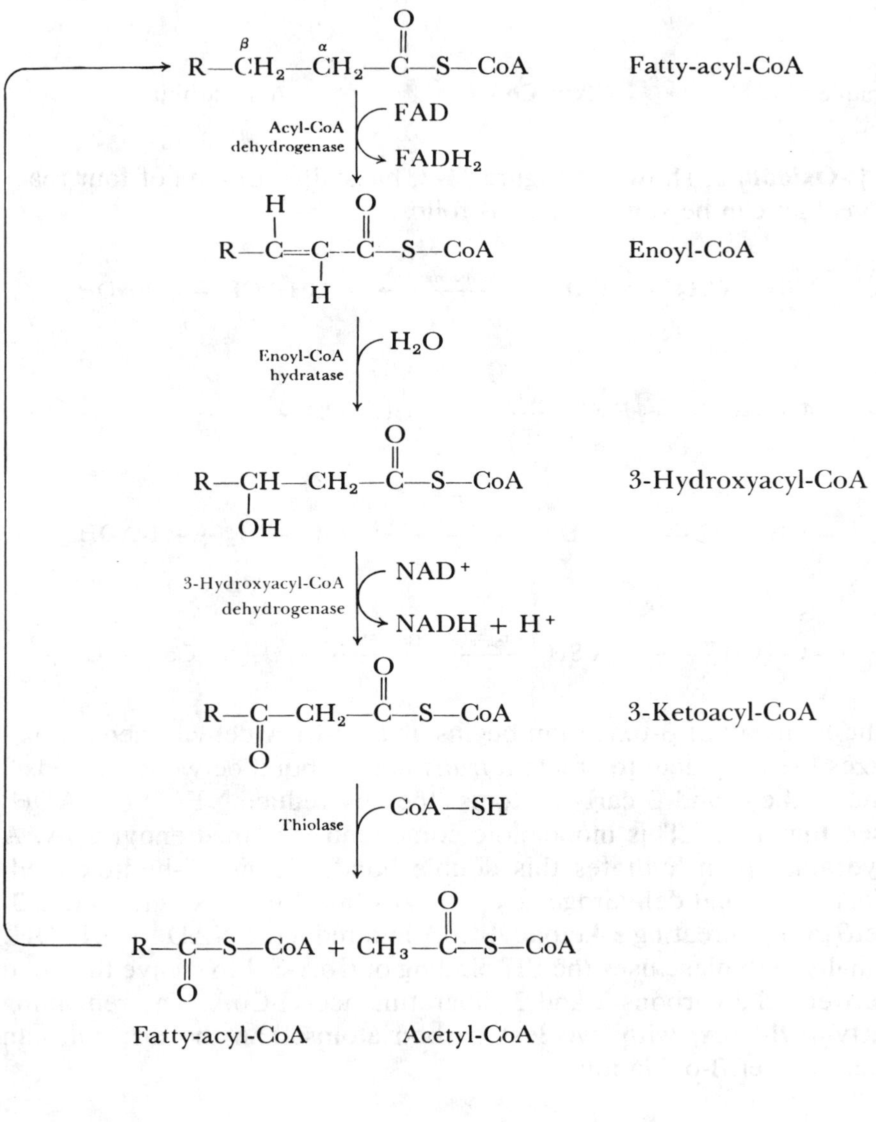

ω-Oxidation of fatty acids is a minor pathway that is found in the liver.

Fatty-acyl-CoA formed on the outer mitochondrial membrane cannot itself penetrate the mitochondrial membrane. Fatty acids enter the mitochondria only after binding to carnitine to produce fatty-acyl-carnitine, as shown on the bottom of Figure 11-2. The enzyme carnitine acyltransferase I, on the outer surface of the mitochondrial membrane, catalyzes the formation of fatty-acyl-carnitine. Carnitine translocase then transports this molecule across the mitochondrial membrane. Once inside the mitochondria, a second enzyme, carnitine acyltransferase II, on the inner surface of the mitochondrial membrane, catalyzes the reverse reaction, liberating fatty-acyl-CoA.

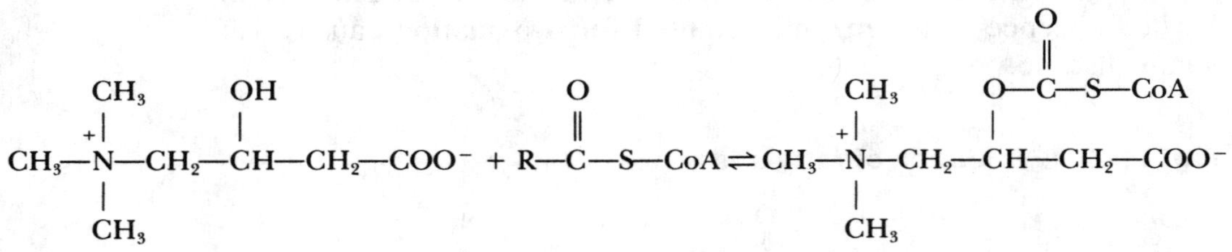

$$CH_3\text{—}\overset{\overset{CH_3}{|}}{\underset{\underset{CH_3}{|}}{\overset{+}{N}}}\text{—}CH_2\text{—}\overset{\overset{OH}{|}}{CH}\text{—}CH_2\text{—}COO^- + R\text{—}\overset{\overset{O}{\|}}{C}\text{—}S\text{—}CoA \rightleftharpoons CH_3\text{—}\overset{\overset{CH_3}{|}}{\underset{\underset{CH_3}{|}}{\overset{+}{N}}}\text{—}CH_2\text{—}\overset{\overset{O\text{—}\overset{\overset{O}{\|}}{C}\text{—}S\text{—}CoA}{|}}{CH}\text{—}CH_2\text{—}COO^-$$

| Carnitine | Acyl-CoA | Acyl-carnitine |

β-Oxidation, shown in Figure 11-2, basically consists of four reactions that can be summarized as follows:

1. $\text{—}\overset{\beta}{C}H_2\text{—}\overset{\alpha}{C}H_2\text{—} + FAD \xrightarrow{\text{Dehydrogenation}} \text{—}CH\text{=}CH\text{—} + FADH_2$

2. $\text{—}CH\text{=}CH\text{—} + H_2O \xrightarrow{\text{Hydration}} \text{—}\overset{\overset{OH}{|}}{C}H\text{—}CH_2\text{—}$

3. $\text{—}\overset{\overset{OH}{|}}{C}H\text{—}CH_2\text{—} + NAD^+ \xrightarrow{\text{Dehydrogenation}} \text{—}\overset{\overset{O}{\|}}{C}\text{—}CH_2\text{—} + NADH + H^+$

4. $\text{—}\overset{\overset{O}{\|}}{C}\text{—}CH_2\text{—} + CoA\text{-}SH \xrightarrow{\text{Cleavage by CoA-SH}} \text{—}\overset{\overset{O}{\|}}{C}\text{—}S\text{—}CoA + CH_3\text{—}$

The pathway of β-oxidation begins as acyl-CoA dehydrogenase oxidizes the fatty acid to create a *trans* double bond between carbons 2 and 3, the α and β carbon atoms, thereby reducing FAD to FADH₂ (see Fig. 11-2). This monoenoic compound is named enoyl-CoA. A hydratase then hydrates this double bond, yielding 3-hydroxyacyl-CoA. A second dehydrogenase oxidizes this 3-hydroxy group to a 3-keto group, creating 3-ketoacyl-CoA and reducing NAD⁺ to NADH. Finally, a thiolase uses the SH binding of CoA-SH to cleave the bond between the carbons 2 and 3, liberating acetyl-CoA. The remaining fatty-acyl-CoA, with two less carbon atoms than the original, can then reenter β-oxidation.

Each pass through β-oxidation removes two carbons from the fatty acid as acetyl-CoA and produces 1 mole $FADH_2$ and 1 mole NADH, which yield 5 moles ATP after reoxidation to FAD and NAD^+ in the electron transport chain (2 moles ATP per mole of $FADH_2$ oxidized and 3 moles ATP per mole of NADH oxidized). The complete β-oxidation of a 12-carbon, saturated fatty acid, for example, requires five passes through this sequence; hence, 6 moles acetyl-CoA, 5 moles $FADH_2$, and 5 moles NADH are produced. Five moles of O_2 are utilized by electron transport to oxidize these $FADH_2$ and NADH and yield 25 moles ATP. Since 2 moles of ATP were consumed to activate the fatty acid—i.e., to convert it to fatty-acyl-CoA—the net ATP yield is 23 moles. If the 6 moles of acetyl-CoA produced by β-oxidation are oxidized in the TCA cycle to 12 moles CO_2, then 6 moles GTP, 6 moles $FADH_2$, and 18 moles NADH are generated. Twelve moles of O_2 are utilized to oxidize these $FADH_2$ and NADH, and thus 72 moles ATP are generated from the total oxidation of these six acetyl-CoA molecules. Therefore, a total of 95 moles ATP and 12 moles CO_2 is produced, while 17 moles O_2 are consumed.

The **respiratory quotient (RQ)** for a catabolized substance is defined as the moles of CO_2 produced divided by the moles of O_2 consumed. For this hypothetical 12-carbon, saturated fatty acid, then, the RQ is:

$$RQ = \frac{12\ CO_2\ produced}{17\ O_2\ consumed} = 0.71$$

The complete oxidation of a 12-carbon disaccharide using glycolysis and the TCA cycle produces only 72 or 76 moles ATP, compared to the 96 moles ATP gained by oxidizing a 12-carbon fatty acid. For this disaccharide, since 12 moles CO_2 are produced while 12 moles O_2 are consumed, its RQ is 1.0.

Biologists calculate the RQ for animals to determine their chief metabolic fuel; an RQ near 1.0 denotes reliance on glycolysis, whereas an RQ near 0.7 indicates a reliance on β-oxidation of fatty acids. When the Caloric values of food are calculated, triglyceride is found to have 9 Cal/gram compared to only 4 Cal/gram for carbohydrate and protein. This explains why mammals rely on triglyceride rather than carbohydrate for energy storage.

Catabolism of Odd-Carbon Fatty Acids

Most naturally occurring fatty acids have an even number of carbon atoms and are therefore completely degraded to acetyl-CoA during β-oxidation. However, propionate, a three-carbon fatty acid, arises from the β-oxidation of odd-carbon fatty acids and from the catabolism of two branched-chain amino acids, isoleucine and valine.

In propionate metabolism, a carboxylase adds CO_2 to propionyl-CoA to create methylmalonyl-CoA. Methylmalonyl-CoA mutase, using vitamin B_{12} as its coenzyme, then transfers the

—C(=O)—S—CoA group to the methyl group, yielding succinyl-CoA:

$$
\begin{array}{l}
CH_3 \\
| \\
CH_2 \\
| \\
C=O \\
| \\
S—CoA
\end{array}
+ CO_2
\quad \xrightarrow[\text{ATP}\quad\text{AMP + PP}_i]{}
\quad
\begin{array}{l}
CH_3 \\
| \\
CH—COO^- \\
| \\
C=O \\
| \\
S—CoA
\end{array}
\rightleftharpoons
\begin{array}{l}
\overset{O}{\overset{\|}{CH_2—C—S—CoA}} \\
| \\
CH_2—COO^-
\end{array}
$$

Propionyl-CoA   Methylmalonyl-CoA   Succinyl-CoA

After conversion to oxaloacetate, gluconeogenesis can occur. Thus, for every two odd carbon fatty acids catabolized, one glucose may be created.

Synthesis of Saturated Fatty Acids in the Cytosol

In humans, saturated fatty acids are synthesized from bicarbonate and acetyl-CoA. This cytosolic biosynthesis requires NADPH, which is supplied chiefly by the pentose phosphate shunt. All enzymes of fatty acid biosynthesis are joined in a large multienzyme complex. Acyl-carrier protein (ACP) is employed to carry the elongating fatty acid chain. Like CoA, ACP contains the B vitamin, pantothenic acid.

The first step toward this synthesis is the carboxylation of acetyl-CoA by acetyl-CoA carboxylase and biotin to produce malonyl-CoA. Malonate is a three-carbon, dicarboxylic acid.

$$
CH_3—\overset{O}{\overset{\|}{C}}—S—CoA + HCO_3^- + ATP \xrightleftharpoons[]{\text{Acetyl-CoA}\atop\text{carboxylase}}
$$

$$
{}^-OOC—CH_2—\overset{O}{\overset{\|}{C}}—S—CoA + ADP + P_i
$$

Acetyl-CoA

Malonyl-CoA

Acetyl-CoA carboxylase is an allosteric enzyme activated by polymerization. Citrate is the chief allosteric activator; it promotes polymerization of the enzyme. Thus, when citrate abounds, the cell is rich in energy and can afford to synthesize fatty acids. Palmitoyl-CoA, an endproduct of fatty acid synthesis, reduces polymerization, thereby inhibiting enzyme activity.

Acetyl-CoA is formed in the mitochondria during β-oxidation of fatty acids and the pyruvate-dehydrogenase reaction. Acetyl-CoA formed in mitochondria cannot, however, cross into the cytosol. To supply acetyl-CoA for fatty acid biosynthesis in the cytosol, citrate is used as a shuttle. Citrate formed from acetyl-CoA and OAA in the citrate synthase reaction can diffuse into the cytosol, where it is cleaved by citrate lyase to form acetyl-CoA and OAA. OAA is then converted to malate by malate dehydrogenase. Malate, in turn, is converted to pyruvate by the malic enzyme. Pyruvate is free to enter mitochondria. Once inside, pyruvate is converted to OAA by pyruvate carboxylase, thereby completing the cycle.

The next step in fatty acid biosynthesis involves transacylation of acetyl-CoA and malonyl-CoA by acetyl transacylase and malonyl transacylase.

$$\text{Acetyl-CoA} + \text{ACP} \underset{}{\overset{\text{Acetyl transacylase}}{\rightleftharpoons}} \text{acetyl-ACP} + \text{CoA}$$

$$\text{Malonyl-CoA} + \text{ACP} \overset{\text{Malonyl transacylase}}{\rightleftharpoons} \text{malonyl-ACP} + \text{CoA}$$

As shown in Figure 11-3, β-ketoacyl-ACP synthase then condenses acetyl-ACP and malonyl-ACP, forming acetoacetyl-ACP. During this condensation, malonate liberates its carboxyl group as CO_2.

The remaining reactions of fatty acid biosynthesis accomplish the reverse of the β-oxidation sequence, though using different enzymes. β-Ketoacyl-ACP reductase uses NADPH to reduce the β-keto group of acetoacetyl-ACP to a β-hydroxyl group, giving rise to 3-hydroxylbutyryl-ACP. A dehydratase removes this β-hydroxyl group along with a hydrogen atom, creating the carbon-carbon double bond of

Fig. 11-3 Cytoplasmic fatty acid biosynthesis.

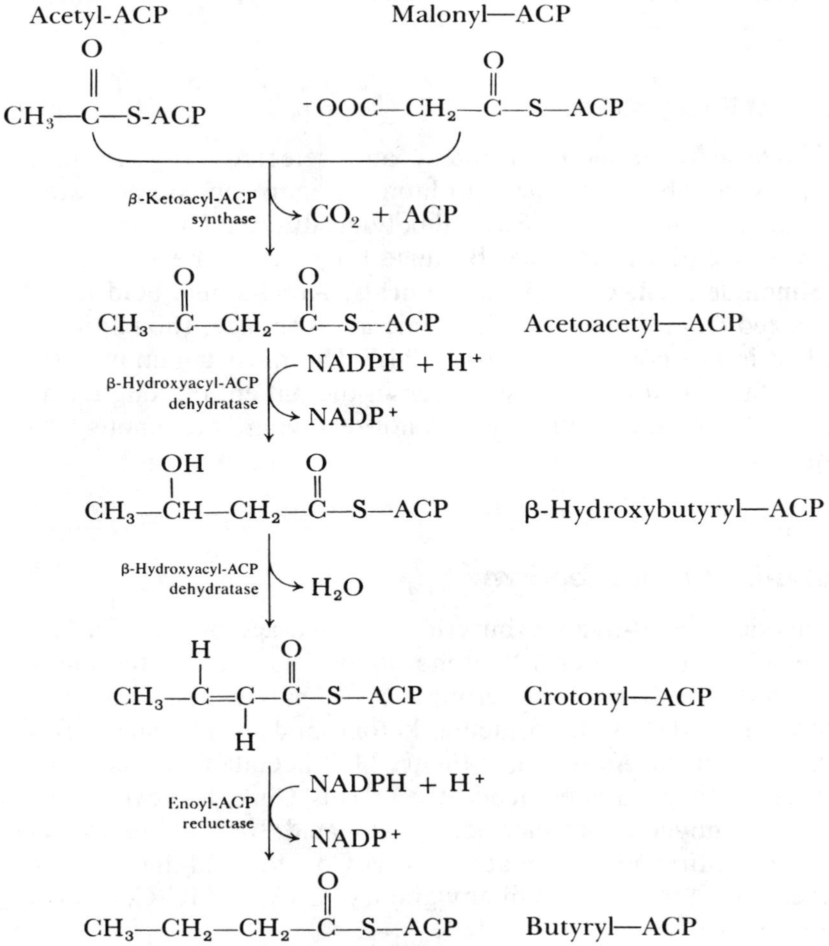

crotonyl-ACP. Enoyl-ACP reductase then saturates this double bond using NADPH to yield butyryl-ACP.

The cycle then repeats itself as butyryl-ACP condenses with malonyl-ACP. Ultimately, a 16-carbon compound, palmitoyl-ACP, is generated, and a thioesterase cleaves this to palmitic acid plus ACP.

For each turn of the cycle, 2 moles of NADPH are consumed by the two reductase reactions, and 1 mole of ATP is consumed in synthesizing malonyl-CoA. Since each mole of NADPH could yield 3 moles ATP in oxidative phosphorylation, each turn of the cycle consumes 7 moles ATP.

To synthesize fatty acids with more than 16 carbon atoms, the palmitic acid from this biosynthesis must enter a fatty acid elongation pathway in the microsomes on the endoplasmic reticulum.

Fatty Acid Elongation and Unsaturation

Saturated and unsaturated fatty acids are lengthened by two-carbon increments by the addition of malonyl-CoA in the microsomes. One mole of CO_2 is liberated during each condensation.

Microsomes use NADH-cytochrome b_5 reductase, cytochrome b_5, and desaturase to create the double bonds of monounsaturated fatty acids such as oleic and palmitoleic acids.

Essential Fatty Acids

Since saturated, monounsaturated, and selected polyunsaturated fatty acids may be synthesized in humans, man can compensate for their absence from the diet. Several polyunsaturated fatty acids, however, are essential in the diet because they cannot be synthesized. These include **linoleic** and **linolenic acids.** Arachidonic acid may be synthesized from linoleic and linolenic acids; hence, though it is required, it is not considered "essential." The most common case of essential fatty acid deficiency is seen in the patient on long-term intravenous hyperalimentation who is not receiving intravenous fat solutions.

Ketone-Body Metabolism

Acetoacetic acid, β-hydroxybutyric acid, and acetone are classified as **ketone bodies.** The term "ketone bodies" is inaccurate, since β-hydroxybutyrate lacks a keto group.

Acetoacetic acid is the principal ketone body synthesized by the liver mitochondria. As shown in Figure 11-4, acetate from acetyl-CoA is dimerized to yield acetoacetyl-CoA. This CoA, however, cannot be readily removed to produce acetoacetic acid. Instead, another acetyl-CoA must first add to the acetoacetyl-CoA to yield the six-carbon intermediate, β-hydroxy-β-methylglutaryl-CoA (HMG-CoA). Ace-

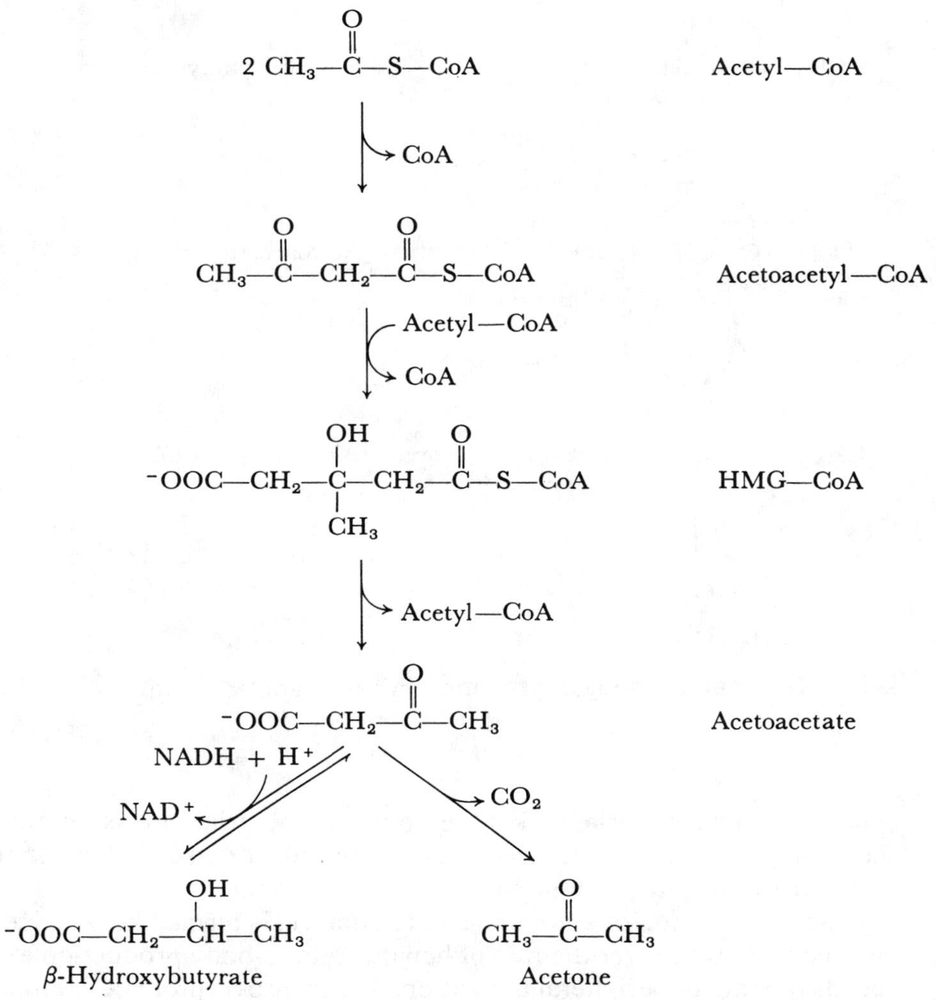

Fig. 11-4 Ketone-body synthesis in the liver.

tyl-CoA is then removed from HMG-CoA in the liver to liberate acetoacetic acid. β-Hydroxybutyrate dehydrogenase reduces much of this acetoacetic acid to β-hydroxybutyric acid. In addition, a decarboxylase converts some of this acetoacetate to acetone. Acetone is metabolized very slowly. Because of its volatility, most of the acetone evaporates through the lung alveoli.

The liver produces ketone bodies when the rate of acetyl-CoA formation exceeds that of acetyl-CoA utilization by the citric acid cycle. Extrahepatic tissues, such as skeletal muscle and heart muscle, utilize the two ketone bodies other than acetone as a fuel (acetone cannot be significantly degraded). These tissues oxidize β-hydroxybutyrate to acetoacetate and then add CoA by either of two different routes to create acetoacetyl-CoA, as shown in Figure 11-5. Finally, they cleave acetoacetyl-CoA into two acetyl-CoA molecules, which can enter the TCA cycle.

High serum levels of acetoacetate and β-hydroxybutyrate constitute **ketonemia.** Normal urine lacks ketone bodies. The common hos-

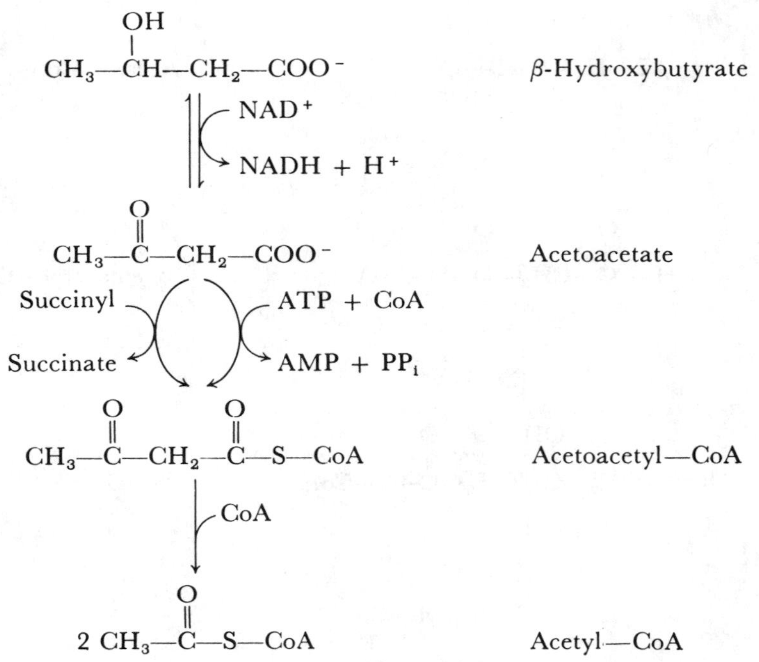

$$
\begin{array}{l}
\overset{\displaystyle OH}{\underset{\displaystyle |}{}} \\
CH_3-CH-CH_2-COO^-
\end{array}
\qquad \beta\text{-Hydroxybutyrate}
$$

NAD$^+$

NADH + H$^+$

$$
\begin{array}{l}
\overset{\displaystyle O}{\underset{\displaystyle \|}{}} \\
CH_3-C-CH_2-COO^-
\end{array}
\qquad \text{Acetoacetate}
$$

Succinyl   ATP + CoA

Succinate   AMP + PP$_i$

$$
\begin{array}{l}
\overset{\displaystyle O}{\underset{\displaystyle \|}{}} \qquad \overset{\displaystyle O}{\underset{\displaystyle \|}{}} \\
CH_3-C-CH_2-C-S-CoA
\end{array}
\qquad \text{Acetoacetyl—CoA}
$$

CoA

$$
2\ \begin{array}{l}
\overset{\displaystyle O}{\underset{\displaystyle \|}{}} \\
CH_3-C-S-CoA
\end{array}
\qquad \text{Acetyl—CoA}
$$

Fig. 11-5 Ketone-body degradation in extrahepatic tissues.

pital tests for **ketonuria,** or ketone bodies in the urine, measure only acetone and acetoacetate. Such tests may fail to detect ketonuria if β-hydroxybutyrate predominates.

The combination of ketonemia and ketonuria is termed **ketosis.** Ketosis occurs whenever the rate of hepatic ketone-body production exceeds the rate of peripheral utilization. The liver overproduces ketone bodies during severe carbohydrate deficiency for at least two reasons. First, carbohydrate deficiency depletes the TCA-cycle intermediates and slows the entrance of acetyl-CoA into this cycle. Second, acetyl-CoA carboxylase, the rate-controlling enzyme of fatty-acid synthesis, is inhibited by the absence of citrate, thereby blocking another route of acetyl-CoA metabolism. Thus, acetyl-CoA accumulates in the liver and is excessively converted to ketone bodies. Severe ketonemia overwhelms the blood buffers, causing metabolic acidosis with an increased anion gap (the "unmeasured anions" in this case are β-hydroxybutyrate and acetoacetate). The high glucagon/insulin ratio during severe carbohydrate deficiency is the mechanism underlying diabetic ketoacidosis, alcoholic ketoacidosis, and starvation ketosis.

The Food and Nutrition Board of the United States recommends that the adult diet contain at least 100 g or 400 Cal of carbohydrate daily to generate enough oxaloacetate to maintain the TCA cycle and prevent ketosis. In addition, carbohydrate deficiency causes protein wasting, because much of the dietary amino acids are converted via deamination and gluconeogenesis to glucose.

Cholesterol Biosynthesis

Humans have two sources of cholesterol: dietary cholesterol and its de novo synthesis from acetate. The greater the dietary intake of cholesterol, the lower the rate of cholesterol biosynthesis in the liver and adrenal cortex. Although cholesterol itself has no caloric value in foods, its presence in the diet spares the energy needed to synthesize cholesterol.

The first two steps of cholesterol biosynthesis are the same as those in ketone-body synthesis; acetyl-CoA is dimerized to acetoacetyl-CoA, to which another acetyl-CoA is added to create β-hydroxy-β-methylglutaryl-CoA (HMG-CoA). HMG-CoA formed in the mitochondria is used to form ketone bodies, whereas HMG-CoA formed in the cytoplasm is used for cholesterol synthesis.

HMG-CoA reductase then catalyzes the committed and rate-controlling step of cholesterol biosynthesis, the reduction of the —C(=O)—S—CoA in HMG-CoA to —CH$_2$OH in mevalonic acid, as shown in Figure 11-6. Fasting reduces cholesterol synthesis by reducing the synthesis of this reductase enzyme. Compactin and mevinolin resemble HMG-CoA and competitively inhibit HMG-CoA reductase, leading to a marked drop in serum cholesterol.

Next, mevalonic acid is phosphorylated three times with ATP to create mevalonate-3-phospho-5-pyrophosphate, which loses its carboxyl and 5-pyrophosphate groups to become the five-carbon intermediate, isopentenyl pyrophosphate. An isomerase changes the position of the carbon-carbon double bond in this intermediate to yield another five-carbon intermediate, 3,3-dimethylallyl pyrophosphate.

These two five-carbon intermediates then condense to create the 10-carbon compound, geranyl pyrophosphate. To this, another five-carbon isopentenyl pyrophosphate is added to produce the 15-carbon compound, farnesyl pyrophosphate, as shown in Figure 11-7.

Two farnesyl pyrophosphate molecules dimerize to form presqualene pyrophosphate, which is then reduced to the 30-carbon compound, squalene. Oxygenation is followed by cyclization to yield lanosterol, the first sterol of the pathway. Three methyl groups are subsequently removed to shape this sterol into the 27-carbon product, cholesterol.

The principal cause of cholesterol gallstones is an increase in the biliary ratio of cholesterol to bile salts. Bile salts and phosphoglycerides normally solubilize biliary cholesterol. In bile-salt deficiency, this cholesterol precipitates.

Most circulating cholesterol exists as an ester with a fatty acid. Phosphatidylcholine donates this fatty acid via the enzyme lecithin-cholesterol acyltransferase (LCAT).

$$\text{Cholesterol + lecithin} \xrightarrow{\text{LCAT}} \text{cholesterol ester + lysolecithin}$$

In the liver, fatty actyl-CoA donates a fatty acid to esterify cholesterol.

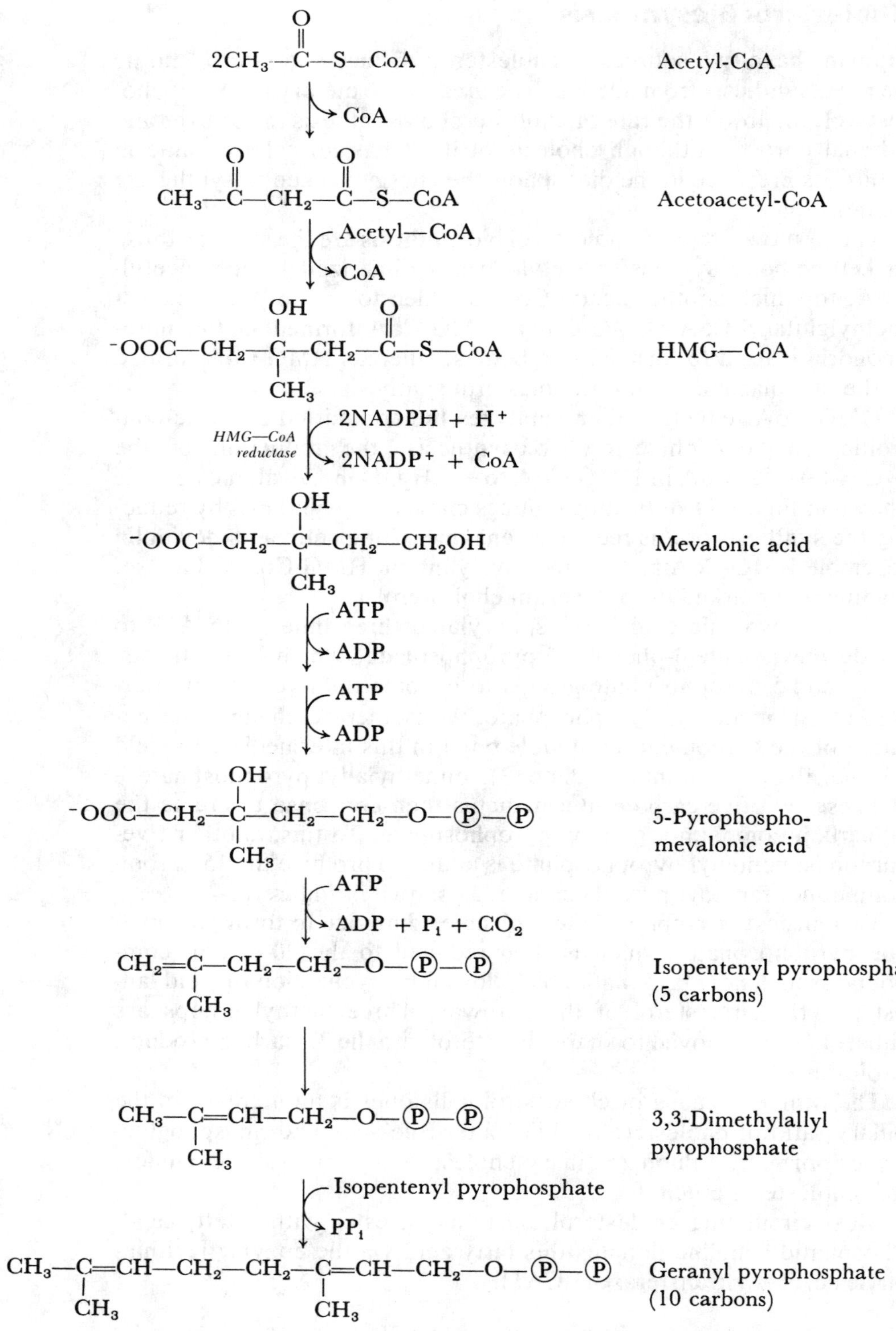

$$2CH_3-\overset{\overset{\displaystyle O}{\|}}{C}-S-CoA \qquad \text{Acetyl-CoA}$$

$$\searrow CoA$$

$$CH_3-\overset{\overset{\displaystyle O}{\|}}{C}-CH_2-\overset{\overset{\displaystyle O}{\|}}{C}-S-CoA \qquad \text{Acetoacetyl-CoA}$$

$$\nearrow \text{Acetyl—CoA}$$
$$\searrow CoA$$

$$^-OOC-CH_2-\underset{\underset{\displaystyle CH_3}{|}}{\overset{\overset{\displaystyle OH}{|}}{C}}-CH_2-\overset{\overset{\displaystyle O}{\|}}{C}-S-CoA \qquad \text{HMG—CoA}$$

$$\underset{\text{reductase}}{HMG-CoA} \nearrow 2NADPH + H^+$$
$$\searrow 2NADP^+ + CoA$$

$$^-OOC-CH_2-\underset{\underset{\displaystyle CH_3}{|}}{\overset{\overset{\displaystyle OH}{|}}{C}}-CH_2-CH_2OH \qquad \text{Mevalonic acid}$$

$$\nearrow ATP$$
$$\searrow ADP$$

$$\nearrow ATP$$
$$\searrow ADP$$

$$^-OOC-CH_2-\underset{\underset{\displaystyle CH_3}{|}}{\overset{\overset{\displaystyle OH}{|}}{C}}-CH_2-CH_2-O-\textcircled{P}-\textcircled{P} \qquad \begin{array}{l}\text{5-Pyrophospho-}\\\text{mevalonic acid}\end{array}$$

$$\nearrow ATP$$
$$\searrow ADP + P_i + CO_2$$

$$CH_2{=}\underset{\underset{\displaystyle CH_3}{|}}{C}-CH_2-CH_2-O-\textcircled{P}-\textcircled{P} \qquad \begin{array}{l}\text{Isopentenyl pyrophosphate}\\\text{(5 carbons)}\end{array}$$

$$\downarrow$$

$$CH_3-\underset{\underset{\displaystyle CH_3}{|}}{C}{=}CH-CH_2-O-\textcircled{P}-\textcircled{P} \qquad \begin{array}{l}\text{3,3-Dimethylallyl}\\\text{pyrophosphate}\end{array}$$

$$\nearrow \text{Isopentenyl pyrophosphate}$$
$$\searrow PP_i$$

$$CH_3-\underset{\underset{\displaystyle CH_3}{|}}{C}{=}CH-CH_2-CH_2-\underset{\underset{\displaystyle CH_3}{|}}{C}{=}CH-CH_2-O-\textcircled{P}-\textcircled{P} \qquad \begin{array}{l}\text{Geranyl pyrophosphate}\\\text{(10 carbons)}\end{array}$$

Fig. 11-6 Cholesterol biosynthetic pathway from acetyl-CoA to geranyl pyrophosphate.

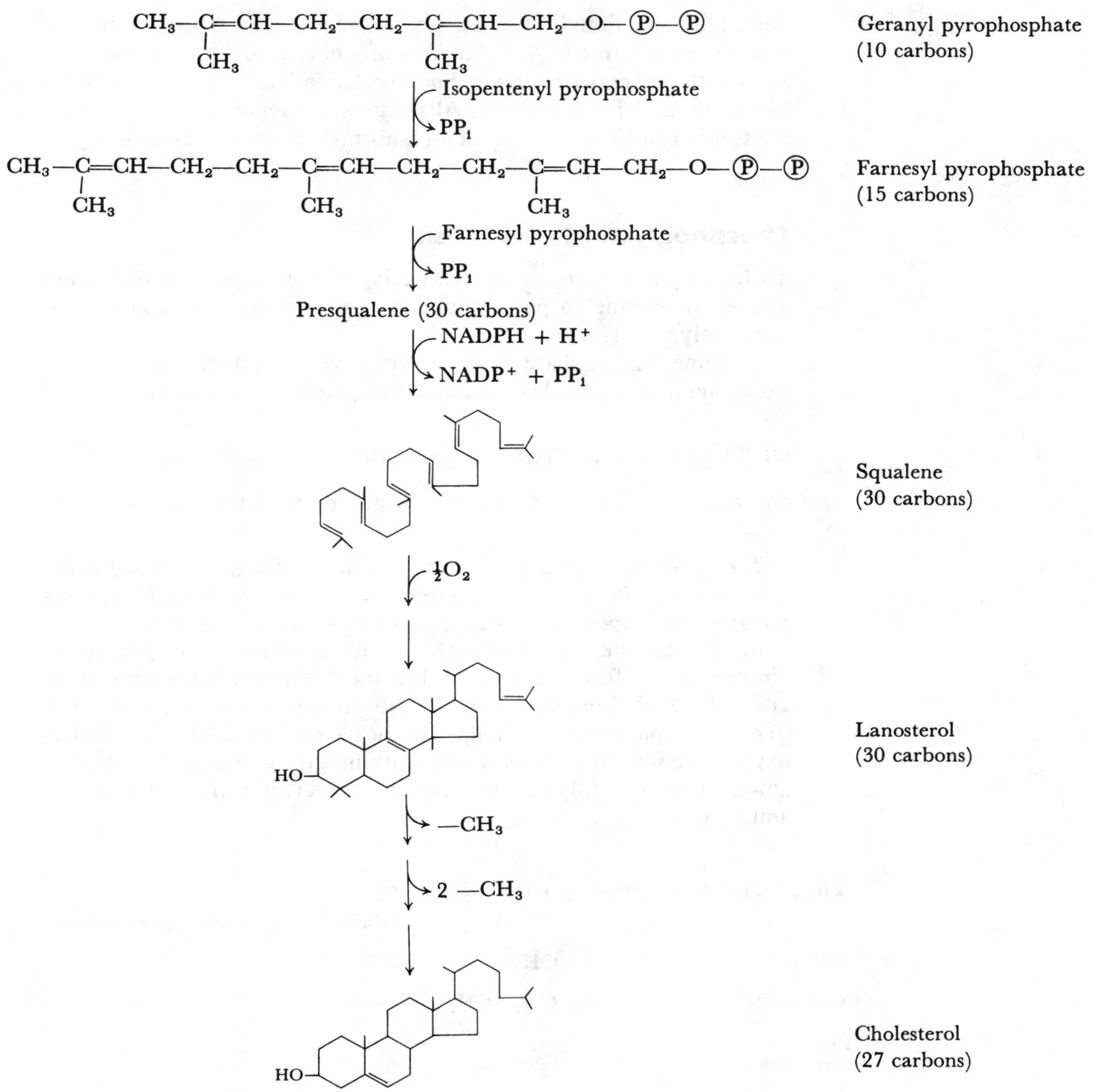

$$CH_3-\underset{\underset{CH_3}{|}}{C}=CH-CH_2-CH_2-\underset{\underset{CH_3}{|}}{C}=CH-CH_2-O-\textcircled{P}-\textcircled{P}$$

Geranyl pyrophosphate
(10 carbons)

Isopentenyl pyrophosphate

PP$_i$

$$CH_3-\underset{\underset{CH_3}{|}}{C}=CH-CH_2-CH_2-\underset{\underset{CH_3}{|}}{C}=CH-CH_2-CH_2-\underset{\underset{CH_3}{|}}{C}=CH-CH_2-O-\textcircled{P}-\textcircled{P}$$

Farnesyl pyrophosphate
(15 carbons)

Farnesyl pyrophosphate

PP$_i$

Presqualene (30 carbons)

NADPH + H$^+$

NADP$^+$ + PP$_i$

Squalene
(30 carbons)

$\frac{1}{2}O_2$

Lanosterol
(30 carbons)

—CH$_3$

2 —CH$_3$

Cholesterol
(27 carbons)

Fig. 11-7 Cholesterol biosynthetic pathway from geranyl pyrophosphate to cholesterol.

Hyperlipoproteinemias

Most of the hyperlipoproteinemias are acquired or sporadic rather than familial. Their common causes include diabetes mellitus, hypothyroidism, and high saturated fat intake.

With normal cells, LDL in high concentrations saturates the LDL membrane receptors, which signals the cell to stop synthesizing cholesterol. Those rare individuals with **homozygous hypercholesterol-**

emia lack functional LDL membrane receptors, so that even high levels of circulating LDL fail to shut off cholesterol biosynthesis. Persons with **heterozygous hypercholesterolemia** have a reduced number of functional LDL receptors. Although some feedback inhibition occurs, it is insufficient to maintain a normal serum cholesterol level.

Phosphoglyceride Synthesis

To begin phosphoglyceride synthesis, ATP phosphorylates choline and ethanolamine to phosphocholine and phosphoethanolamine respectively.

Cytidine triphosphate (CTP) is employed to attach phosphoethanolamine to diglycerides to create phosphatidylethanolamine:

$$\text{CTP} + \text{phosphoethanolamine} \rightleftharpoons \text{CDP-ethanolamine} + \text{PP}_i$$

$$\text{CDP-ethanolamine} + \text{diglyceride} \rightleftharpoons \text{phosphatidylethanolamine} + \text{CMP}$$

The synthesis of phosphatidylcholine in the lungs of newborns is a critical factor in enabling the postnatal expansion of the lung, since phosphatidylcholine serves as a pulmonary surfactant.

In humans, there are two pathways for synthesizing phosphatidylcholine. In the first pathway, CTP combines with phosphocholine to create CDP-choline, which donates its choline group to a diglyceride to generate phosphatidylcholine. In the second, each of three *S*-adenosylmethionine molecules donates its methyl group to the ethanolamine of phosphatidylethanolamine to convert it to trimethylethanolamine, or choline:

$$\text{Phosphatidylethanolamine} + 3 \; S\text{-adenosylmethionine} \longrightarrow$$
$$\text{phosphatidylcholine} + 3 \; S\text{-adenosylhomocysteine}$$

$$\begin{array}{ll}
\overset{\displaystyle \text{OH}}{\underset{\displaystyle |}{}} & \overset{\displaystyle \text{OH}}{\underset{\displaystyle |}{}} \qquad \overset{\displaystyle \text{CH}_3}{\underset{\displaystyle |}{}} \\
\text{CH}_2\text{—CH}_2\text{—}^+\text{NH}_3 & \text{CH}_2\text{—CH}_2\text{—}^+\text{N—CH}_3 \\
& \qquad\qquad\quad | \\
& \qquad\qquad\; \text{CH}_3 \\
\text{Ethanolamine} & \text{Choline}
\end{array}$$

Sphingolipid Catabolism

A variety of hydrolytic enzymes in lysosomes are required to catabolize sphingolipids. The absence of any one of these enzymes leads to a sphingolipid-deposition disease known as a **sphingolipidosis.**

In **Niemann-Pick disease,** the lack of **sphingomyelinase** prevents the removal of phosphocholine from ceramide in sphingomyelin. Thus, sphingomyelin accumulates within the brain, liver, and spleen.

Absence of **hexosaminidase A** in **Tay-Sachs disease** blocks the cleavage of *N*-acetylgalactosamine (a hexosamine) from GM$_2$ gangliosides, which deposit in the brain and in the macula of the retina, causing cherry-red spots.

In **metachromatic leukodystrophy,** the absence of a **sulfatidase,** arylsulfatase A, permits sulfatides to accumulate in the white matter of the brain.

β-Glucosidase deficiency in **Gaucher's disease** prevents the cleavage of glucose from glucocerebrosides, which deposit in the liver, the spleen, and, in infants, the brain. The bone marrow exhibits cerebroside-filled Gaucher cells.

Absence of **α-galactosidase** in **Fabry's disease** causes ceramide-trisaccharides to accumulate in the skin and kidneys.

Hormonal Control of Lipid Metabolism

Insulin promotes triglyceride synthesis by three mechanisms. First and foremost, it hastens glucose entry into adipose tissue, thereby providing fuel for fatty-acid and triglyceride synthesis. Second, insulin stimulates enzymes involved in fatty-acid synthesis in adipose tissue and liver. Finally, insulin inhibits the hormone-sensitive lipase in adipose tissue by promoting its dephosphorylation, which renders the lipase inactive. Thus, insulin is an **antilipolytic** hormone; somatomedins behave similarly.

Lipolytic hormones stimulate adenyl cyclase in adipose tissues to convert ATP to cyclic AMP. Cyclic AMP activates protein kinase A, which in turn phosphorylates hormone-sensitive lipase. This activated lipase hydrolyzes triglycerides and releases a surge of albumin-bound, fatty acids into the blood. This cyclic-AMP-mediated control of lipolysis in adipose tissue resembles cyclic-AMP-induced glycogenolysis in the liver. The activated protein kinase A also phosphorylates acetyl-CoA carboxylase, thereby inactivating it and turning off fatty-acid biosynthesis. Hence, increased lipolysis is linked to decreased lipid synthesis.

The most potent lipolytic hormone is epinephrine. The plasma free-fatty-acid level rises markedly after an injection of epinephrine. Other lipolytic hormones include growth hormone (STH), thyroid hormones, and glucagon.

Hyperthyroidism makes the hormone-dependent lipase more sensitive to epinephrine. Hypercholesterolemia (high serum cholesterol levels) occurs in **hypothyroidism,** because of impaired removal of cholesterol from the blood.

Glucocorticoid deficiency in adrenocortical hypofunction reduces the sensitivity of hormone-sensitive lipase to lipolytic agents. **Glucocorticoid excess,** or **Cushing's syndrome,** causes an unexplained redistribution of fat from the extremities to the trunk, face ("moon face"), and lower neck posteriorly ("buffalo hump").

Immediately after a meal, insulin is released and the lipolytic-hormone levels diminish, allowing insulin to promote fatty-acid, triglyceride, and glycogen synthesis. Insulin, therefore, acts as a fuel-storing hormone (as do the somatomedins). Several hours later, insulin begins to disappear from the bloodstream, and the lipolytic, glycogenolytic, and gluconeogenic hormones start to mobilize triglycerides, glycogen, and proteins, respectively. During fasting, these fuel-mobilizing hormones predominate and insulin almost vanishes. The lack of the insulin effect in diabetes mellitus leads to excess lipid mobilization, which results often in hypertriglyceridemia.

Problems

Problem 1

In contrast to fatty-acid biosynthesis in the cytosol, β-oxidation of saturated fatty acids:

  A. Requires ACP to transfer the fatty acid chain.
  B. Is activated by citrate.
  C. Is blocked by HMG-CoA reductase inhibitors.
  D. Generates malonyl-CoA.
  E. Uses $FADH_2$ and FAD^+.

Problem 2

Choose the incorrect statement about ketone bodies.

  A. During prolonged starvation, the brain adapts to using ketone bodies as a fuel.
  B. Excess acetone production during ketoacidosis is not itself highly dangerous.
  C. β-Hydroxybutyrate can be used as a fuel for fatty-acid biosynthesis.
  D. Acetoacetate can be indirectly converted to glucose during prolonged starvation.
  E. Ketone bodies form when the glucagon/insulin ratio is low.

Problem 3

HMG-CoA:

  A. From the cytosol is converted to ketone bodies.
  B. Formation is the rate-controlling step in cholesterol biosynthesis.
  C. From the mitochondria can be used in cholesterol biosynthesis.
  D. Is formed from acetoacetyl-CoA plus acetyl-CoA.
  E. Can be converted to glucose.

Problem 4

An Antarctic explorer decides to use butter (9 Cal/gm) rather than carbohydrate (4 Cal/gm) as his energy source during an expedition. His protein source, dried beef, has virtually no carbohydrate. After two weeks on such a diet, he would be expected to have elevated serum levels of all of the following except:

  A. Lactate.
  B. Fatty acids.

C. Acetoacetate.
D. β-Hydroxybutyrate.
E. Glucagon.

Problems 5–8

Match the reactions of cholesterol biosynthesis below to the descriptions given.

A. Lanosterol ———→ cholesterol
B. HMG-CoA ———→ mevalonic acid + CoA
C. Acetoacetyl-CoA + acetyl-CoA ———→ HMG-CoA + CoA
D. Squalene ———→ lanosterol

5. Sterol ring created from a branched-chain structure.
6. Loss of three methyl groups.
7. Rate-controlling step of cholesterol biosynthesis.
8. This reaction is used in ketone body synthesis.

Problem 9

Choose the incorrect statement about lipolytic hormones.

A. They lead to the phosphorylation of hormone-sensitive lipase by protein kinase.
B. They lower serum cholesterol levels.
C. They activate adenyl cyclase in adipose tissue.
D. Lipolytic hormones such as glucagon inhibit acetyl-CoA carboxylase.
E. They act via cyclic AMP.

Problems 10–12

Match the compounds below to the shunt or carrier functions they perform.

A. Carnitine
B. Citrate
C. ACP
D. Glycerol-P

10. Transfers acetate from mitochondria to cytoplasm.
11. Transfers fatty acids into mitochondria.
12. Transfers NADH from the cytoplasm into the mitochondria.

Answers

1. E.
2. D. β-Hydroxybutyrate is converted to acetyl-CoA, which is required for fatty-acid biosynthesis. Ketone bodies cannot be converted to glucose.
3. D. HMG-CoA from the cytosol is used in cholesterol biosynthesis, whereas mitochondrial HMG-CoA can be used for ketone body formation. The reduction of HMG-CoA is the rate-controlling step in cholesterol biosynthesis.

4. A. Free fatty acids are released into the serum during starvation.
5. D.
6. A.
7. B.
8. C.
9. B.
10. B. Citrate synthase in the mitochondria creates citrate from acetate and OAA. The citrate then crosses into the cytoplasm, where citrate lyase cleaves it to acetyl-CoA.
11. A.
12. D.

References

Devlin, T. M. *Textbook of Biochemistry with Clinical Correlations* (3rd ed.). New York: Wiley-Liss, 1992. Pp. 387–470.

Mathews, C. K., and van Holde, K. E. *Biochemistry*. Redwood City, Calif.: Benjamin/Cummings, 1990. Pp. 571–641.

Murray, R. K., Granner, D. K., Mayes, P. A., and Rodwell, V. W. *Harper's Biochemistry* (22nd ed.). Norwalk, Conn.: Appleton & Lange, 1990. Pp. 199–260.

Stryer, L. *Biochemistry* (3rd ed.). New York: Freeman, 1988. Pp. 469–483, 547–574.

Amino Acid Catabolism and Biosynthesis

Proteolysis

Proteolysis, or the hydrolysis of the peptide bonds of dietary protein, begins in the stomach, where gastric HCl acidifies the food to pH 2 to 3, the optimum pH for the proteolytic enzyme, pepsin. Since HCl in the stomach is too dilute to hydrolyze peptide bonds readily by itself, gastric proteolysis is carried out primarily by pepsin. On entering the intestine, the oligopeptides and polypeptides produced by pepsin digestion encounter the pancreatic proteasese—i.e., trypsin, chymotrypsin, and carboxypeptidase—and are cleaved to a mixture of free amino acids and oligopeptides. Amino acids are absorbed in the small intestine via Na^+-dependent carriers.

The gastric and pancreatic proteases are secreted as zymogens, or inactive precursors, termed pepsinogen, trypsinogen, and chymotrypsinogen. Gastric HCl activates pepsin, and enterokinase, an intestinal enzyme, activates trypsinogen to trypsin. Trypsin, in turn, activates chymotrypsinogen.

The intestinal mucosa contains and also secretes the protease leucine aminopeptidase and the dipeptidases, which further degrade oligopeptides into free amino acids. The small intestine then absorbs these free amino acids by active transport. They enter the hepatic portal-venous system and are carried to the main organ of amino acid metabolism, the liver.

Amino Acid Catabolism

Most amino acids cannot passively cross cell membranes. Instead, they require active transport via the γ-glutamyl cycle. In this cycle, γ-glutamyl transpeptidase in cell membranes transfers the γ-glutamyl residue of glutathione to the extracellular amino acid, creating an γ-glutamyl amino acid, which crosses the cell membrane. Once inside, γ-glutamyl cyclotransferase removes the γ-glutamyl residue, yielding the free amino acid plus the cyclized amino acid, 5-oxoproline. Several additional reactions of this cycle restore glutathione from 5-oxoproline, cysteine, and glycine.

Amino acids in the blood readily penetrate the renal glomeruli, but they do not normally appear in the urine in appreciable amounts, because the renal tubules avidly reabsorb them. In cystinuria, an inherited disorder, the kidney tubules fail to reabsorb the basic amino acids cystine, ornithine, arginine, and lysine (the mnemonic is COAL). In this and other aminoacidurias, the urine drains significant quantities of amino acids from the blood.

The main site of amino acid catabolism is the liver. The kidney is a secondary site of catabolism (i.e., via metabolic conversion, rather than through loss into the glomerular filtrate). Amino acid catabolism generally begins with the removal of the α-amino group, leaving an α-keto acid. The liver converts most of the ammonium (NH_4^+) ions, produced by deamination, to urea, which is secreted into the urine. Some of these ammonium ions are secreted directly into the urine by the kidneys. The carbon skeleton of the α-keto acid is then degraded to useful intermediates or metabolic endproducts, such as oxalate.

Protein catabolism generates 4 Cal/gram, the same energy yield as in carbohydrate breakdown. Dietary protein has a high **specific dynamic action,** or post-absorptive thermogenesis, which is defined as the quantity of heat, relative to the basal state, released during catabolism following the ingestion of food.

Removal of Alpha-Amino Groups

The principal means for removing α-amino groups is a transamination reaction with α-ketoglutarate that yields the corresponding α-keto acid plus glutamate, as shown in Figure 12-1. (Since glutarate sounds like glutamate, it is easy to remember that they are an α-keto acid and α-amino acid pair.)

Pyridoxal phosphate, a member of the vitamin-B_6 group, is an essential coenzyme in transaminations, as well as in many other amino acid reactions. It may accept an amino group to become pyridoxamine phosphate.

After the aminotransferases have gathered most of the α-amino groups into glutamate, the enzyme glutamate dehydrogenase liberates the amino groups as ammonia and regenerates α-ketoglutarate. This is a deamination, rather than a transamination, reaction. Furthermore, it is an oxidative deamination, because the NAD^+ in this reaction is reduced to NADH, which is reoxidized by the electron transport chain. Glutamate dehydrogenase can utilize either NAD^+ or $NADP^+$. It uses NADPH to aminate α-ketoglutarate.

The second method for removing α-amino groups employs oxidative deamination without transamination. This pathway, of minor importance in human metabolism, uses FMN-linked amino acid oxi-

Fig. 12-1 Transamination followed by oxidative deamination.

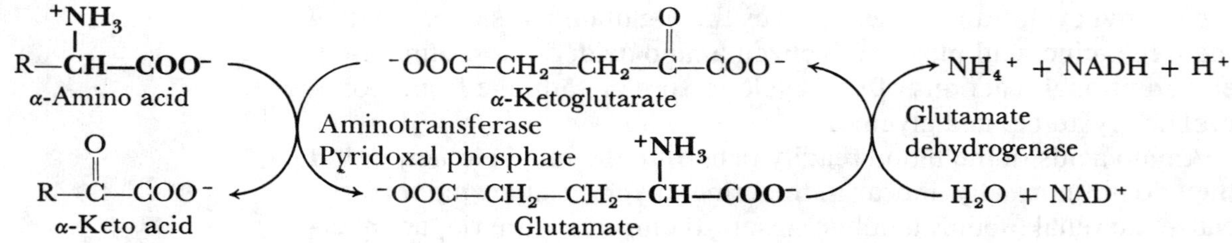

dase. The $FMNH_2$ produced then reduces O_2 to H_2O_2. Human catalase degrades this hydrogen peroxide to $H_2O + \frac{1}{2}O_2$:

$$\alpha\text{-Amino acid} + FMN + H_2O \longrightarrow \alpha\text{-keto acid} + NH_3 + FMNH_2$$

$$FMNH_2 + O_2 \longrightarrow FMN + H_2O_2$$

$$H_2O_2 \xrightarrow{\text{Catalase}} H_2O + \frac{1}{2}O_2$$

The third mechanism for removing α-amino groups is by nonoxidative deamination. This is the main route for removing the α-amino groups from threonine and serine.

Renal Ammonia Excretion

The kidneys excrete 60% of the daily load of hydrogen ions from metabolism as ammonium ions and 40% as titratable acids, chiefly phosphate.

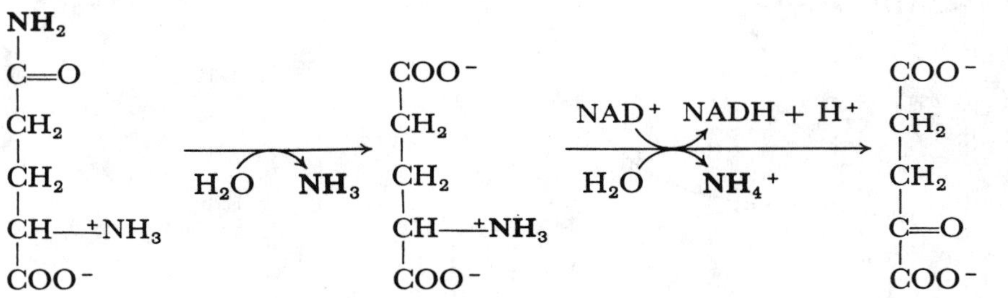

Glutamine   Glutamate α-Ketoglutarate

Glutaminase nonoxidatively deaminates glutamine to glutamate in the kidneys. Glutamate dehydrogenase then oxidatively deaminates glutamate to α-ketoglutarate. Two NH_3 molecules are secreted into the urine for every glutamine twice-deaminated to α-ketoglutarate. Once inside the renal tubules, NH_3 is protonated to NH_4^+, which cannot leave the tubules, because of ion trapping.

A second pathway transaminates (rather than deaminates) glutamine and then deaminates the product to yield α-ketoglutarate plus ammonia.

Urea Cycle

Renal ammonia excretion accounts for only a small fraction of the daily loss of the nitrogen resulting from amino acid catabolism. The vast majority of nitrogen excreted is in the form of urea, which penetrates the renal glomeruli and is lost in the urine. The urea cycle occurs predominately in the liver, although it also exists in the brain and kidneys. Many tissues possess all but one urea-cycle enzyme, and they use this pathway to synthesize arginine. Their lack of arginase, however, prevents urea formation from the arginine.

Carbamoyl-phosphate synthetase within mitochondria generates the unstable intermediate, carbamoyl phosphate, from CO_2, NH_4^+, and H_2O, expending 2 moles of ATP in the reaction. The glutamate-dehydrogenase reaction supplies the required NH_4^+. Another carbamoyl-phosphate synthetase exists in the cytoplasm to generate carbamoyl phosphate for pyrimidine synthesis.

To enter the urea cycle, this carbamoyl group, $—C(=O)—NH_2$, is added by ornithine transcarbamoylase inside the mitochondria, to the basic amino acid, ornithine, to create citrulline, as shown in Figure 12-2. *Citrulline diffuses into the cytosol, which is the reaction site for the remaining three urea-cycle enzymes.*

Argininosuccinate synthetase in the cytoplasm condenses citrulline and aspartate to generate argininosuccinate. In the process, ATP is

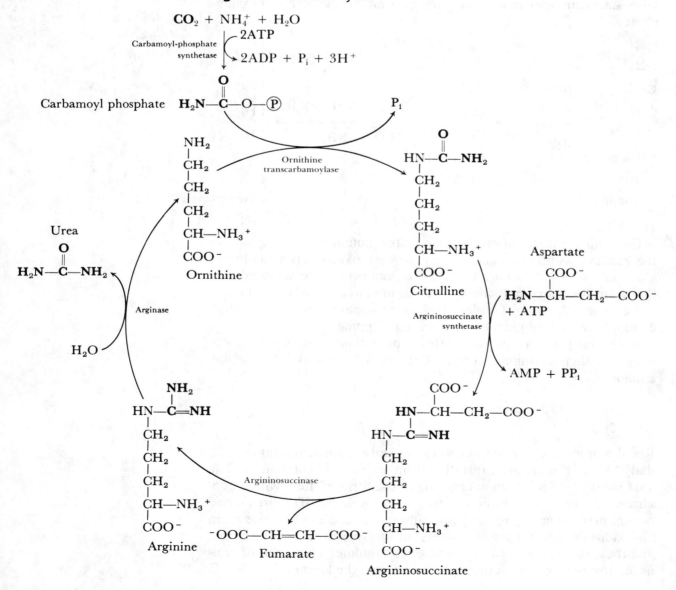

Fig. 12-2   The urea cycle.

split to AMP + PP_i (the pyrophosphate is then hydrolyzed to 2 moles P_i).

Argininosuccinase cleaves argininosuccinate to arginine and fumarate; it transforms a single bond in succinate to the double bond of fumarate. This fumarate enters the TCA cycle.

Many tissues possess the enzymes needed to synthesize arginine by this pathway. Only the liver, kidneys, and brain have the arginase to hydrolyze arginine to urea and ornithine. Ornithine completes the urea cycle by returning to the mitochondria, where it again condenses with carbamoyl phosphate.

One nitrogen atom of the urea produced comes from glutamate via carbamoyl phosphate, while the other is taken from aspartate in the argininosuccinate-synthetase step. The carbon atom of the urea is derived from CO_2, also via carbamoyl phosphate.

The net reaction of the urea cycle consumes the equivalent of 4 moles ATP; 2 moles ATP must be expended to convert the AMP produced in the argininosuccinate-synthetase step to ATP. The reaction may be written as:

$$CO_2 + NH_4^+ + \text{aspartate} + 3ATP + 2H_2O \longrightarrow$$
$$\text{urea} + 2ADP + AMP + 4P_i + \text{fumarate}$$

For each enzyme of the urea cycle, there is a rare, inherited, deficiency disorder that is almost always fatal in infancy. The urea concentration is extremely low in affected individuals.

Catabolism of the Carbon Skeletons

The carbon skeletons of the twenty common amino acids are catabolized to seven compounds. Five of these seven can be converted to glucose: oxaloacetate, pyruvate, α-ketoglutarate, succinyl CoA, and fumarate. Amino acids convertible to these five are termed glucogenic because these metabolites can be converted to glucose in gluconeogenesis.

The remaining two compounds, acetyl-CoA and acetoacetyl-CoA, can be converted to ketone bodies but not to glucose. Amino acids convertible to these two are termed ketogenic. Leucine and lysine are the only purely ketogenic amino acids. Four amino acids are both glucogenic and ketogenic while the remainder are strictly glucogenic.

Learning all the intricate catabolic pathways for all the individual amino acids is a job for a biochemist specializing in amino acid metabolism; medical students need not attempt such a colossal task. Instead, one need only focus on the general features of the medically relevant pathways outlined in this chapter.

Amino Acid Biosynthesis

Bacteria and higher plants can synthesize all twenty common α-amino acids by adding ammonia to carbon skeletons, such as the α-

keto acids, that they fabricate. Mammals, however, cannot manufacture certain α-keto acids that are required in the synthesis of their corresponding amino acid. Any amino acid that humans either cannot synthesize or are unable to manufacture in adequate quantity is termed **essential.**

An essential amino acid must be provided in the diet. The absence of an essential amino acid from the diet generally, but not always, causes a negative **nitrogen balance;** that is, the total nitrogen losses in the urine, feces, and sweat exceed the dietary nitrogen intake. Replacing the deficient essential amino acid promptly restores the positive nitrogen balance that is needed for growth or the zero nitrogen balance that is required for maintenance.

The absence of a nonessential amino acid from the diet, however, does not impair protein synthesis, because such an amino acid can be manufactured in adequate amounts. Essential amino acids are *not* more important in human metabolism than the nonessential amino acids; they differ only in that the essential amino acids must be provided in the diet. Humans could synthesize many, but not all, essential amino acids by transaminating the corresponding α-keto acids, but, since these keto acids cannot be synthesized and do not appear in the diet, these pathways are normally of no avail.

Nine amino acids are essential for humans: histidine, isoleucine, leucine, lysine, methionine, phenylalanine, threonine, tryptophan, and valine. Histidine differs from the other essential amino acids in that its short-term deficiency in humans does *not* produce a negative nitrogen balance. Nevertheless, humans cannot synthesize any histidine. Although humans can synthesize arginine using the urea-cycle enzymes, most of this arginine is degraded to ornithine plus urea. Hence, children may not synthesize enough arginine to meet their needs.

Cysteine and tyrosine are in the gray zone between essential and nonessential. The absence of cysteine from the diet, for example, raises the methionine requirement by 30%, because cysteine is synthesized from methionine. Similarly, the absence of tyrosine increases the phenylalanine requirement. Thus, tyrosine and cysteine are set apart from the other nonessential amino acids. The combined dietary methionine-cysteine and phenylalanine-tyrosine intakes are used to compute the requirements for these amino acids.

The essential amino acids can be classified as follows. An asterisk is used to denote the two amino acids that are nonessential but are synthesized from essential amino acids:

1. All three branched-chain amino acids (isoleucine, leucine, valine).
2. All three aromatic amino acids (phenylalanine-tyrosine,* tryptophan).
3. Both sulfur-containing amino acids (methionine-cysteine*).
4. Two basic amino acids (histidine, lysine).
5. Threonine.

The syntheses of essential amino acids by plants and bacteria are generally longer and more intricate than those for nonessential amino acids.

Each essential amino acid is required in differing amounts. Humans must ingest four times more leucine than tryptophan, for example. A protein of high biologic quality has a relative amino acid composition that is similar to the human requirement. The ideal food protein is egg albumin. Other high-quality protein sources include eggs, meats, fish, and poultry. Milk has moderately high-quality protein. Plant proteins, such as those in corn and wheat, are generally of lower quality. Corn, for instance, has too little tryptophan and lysine. For every gram of egg albumin in the human diet, two or more grams of corn protein would be required to equal the nitrogen-retention provided by egg albumin, the nutritionally ideal protein. Of the amino acids in egg albumin, 40% are essential. Artificial proteins composed entirely of essential amino acids actually have a lower biologic quality than egg albumin, because they overtax the body's ability to synthesize the nonessential amino acids. The judicious combination of proteins of low biologic quality can compensate for the relative lack of certain essential amino acids in each of the component proteins.

Phenylalanine and Tyrosine Metabolism

As shown in Figure 12-3, phenylalanine hydroxylase transforms the essential amino acid phenylalanine into p-hydroxyphenylalanine, or tyrosine. Phenylalanine hydroxylase uses **tetrahydrobiopterin (H_4-Biopterin),** a pteridine compound, as a coenzyme to reduce $NADP^+$ to NADPH. After oxidation to dihydrobiopterin (H_2-Biopterin), NADPH regenerates H_4-Biopterin.

The inherited absence of phenylalanine hydroxylase diverts phenylalanine metabolism into a minor pathway that is not employed normally. Figure 12-3 shows that in this pathway, the alanine portion of the molecule is transaminated to pyruvate, generating phenylpyruvate. This phenylketone spills into the urine; hence the name **phenylketonuria (PKU)** for the disorder resulting from the absence of this enzyme. Phenylpyruvate is reduced to phenyllactate and also decarboxylated to phenylacetate. These aromatic compounds damage the developing brain, producing mental retardation. The brain will develop normally, however, if the phenylalanine intake in the diet is extremely low. An artificial protein that lacks phenylalanine is given to children with PKU to maintain low serum levels of phenylpyruvate and its derivatives. Since they cannot synthesize tyrosine, this becomes an essential amino acid for those with PKU. Whether or not adult phenylketonurics may return to a normal diet without adverse effects is controversial. Certainly, an adult's brain is not as readily damaged by phenylketones as is the infant's developing brain.

In the normal pathway of tyrosine metabolism, tyrosine is transaminated to p-hydroxyphenylpyruvate, which in turn is oxidized to homogentisate by using ascorbate as a coenzyme. This is a typical example of the role of ascorbate in many oxygenation reactions. Homogentisate oxidase then opens the phenyl ring of homogentisate to create a straight-chain compound, which is then cleaved to generate fumarate and acetoacetate. Fumarate enters the TCA cycle,

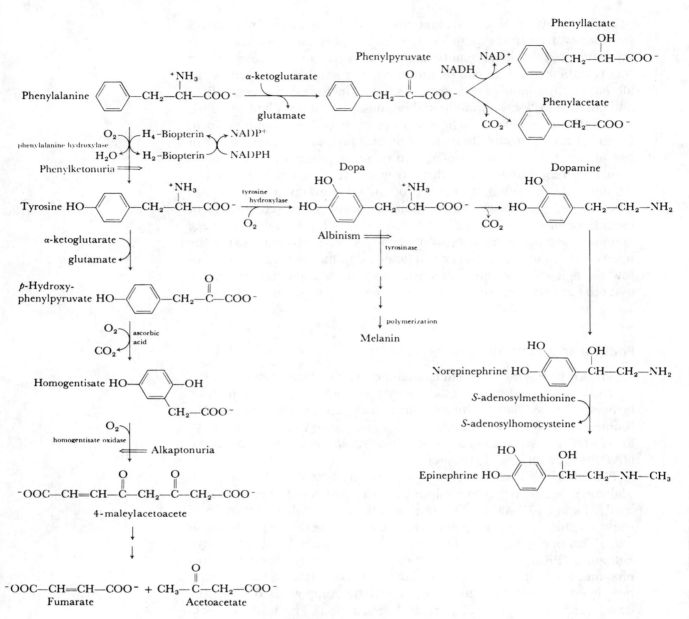

Fig. 12-3 Metabolism of phenylalanine and tyrosine.

thereby rendering phenylalanine and tyrosine glucogenic. They are also ketogenic, because their catabolism yields acetoacetate.

The inherited deficiency of homogentisate oxidase, termed **alkaptonuria,** leads to the buildup of homogentisate derivatives that polymerize into melanin-like pigments. These pigments will stain connective tissue brownish black, and the urine from alkaptonuric patients darkens after standing due to pigment formation.

An alternative route in the normal pathway of tyrosine metabolism is its hydroxylation by tyrosine hydroxylase to *d*ihydroxy*p*henyl-*a*lanine (dopa). This amino acid can then be decarboxylated to dopamine, which is hydroxylated to the neurotransmitter, norepinephrine. *S*-Adenosylmethionine then donates its methyl group to norepinephrine to produce the sympathetic agent, epinephrine. Dopa is administered to patients with Parkinson's disease to correct their lack of dopamine in the basal ganglia of the brain. Another fate of dopa is its hydroxylation by tyrosinase and eventual polymerization to the pigment, melanin. The inherited deficiency of tyrosinase blocks melanin synthesis, which results in albinism.

Tyrosine is iodinated in the thyroid to mono- and diiodotyrosine. Two molecules of diiodotyrosine are joined to create thyroxine (tetraiodothyronine or T_4) and serine. Triiodothyronine (T_3), the most active thyroid hormone, is synthesized by adding mono- to diiodotyrosine or by deiodinating T_4 to T_3.

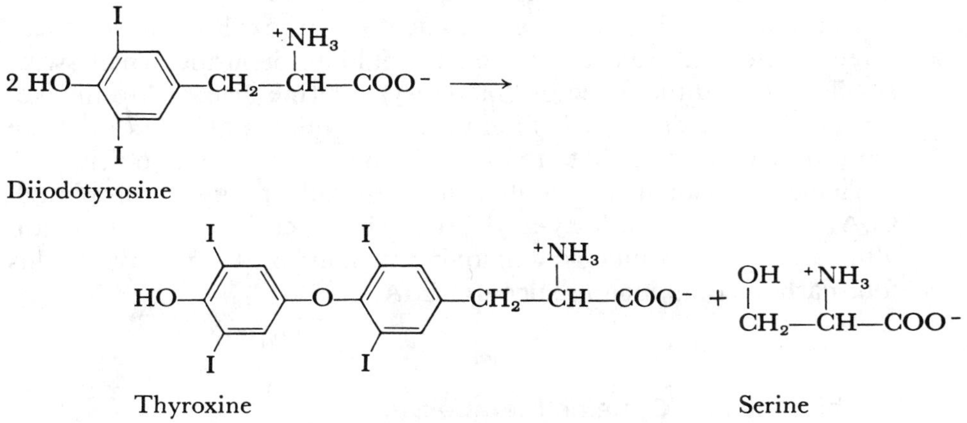

Diiodotyrosine

Thyroxine   Serine

Tryptophan Metabolism

Humans use tryptophan, an essential amino acid, to synthesize part of their nicotinic-acid requirement. Thus, a high tryptophan content in the diet can compensate for a low content of this vitamin. Pellagra classically strikes people subsisting on corn, because corn has little niacin or tryptophan.

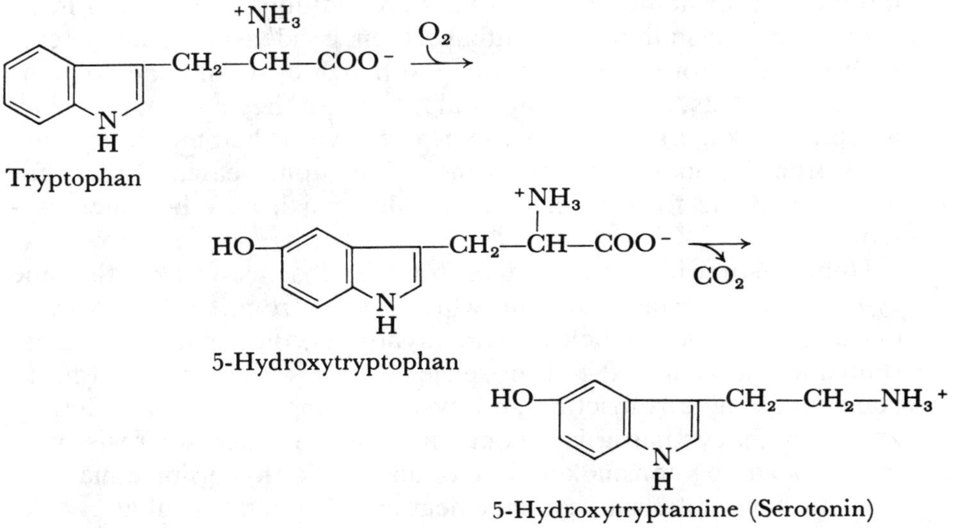

Tryptophan

5-Hydroxytryptophan

5-Hydroxytryptamine (Serotonin)

Tryptophan is also the precursor of 5-hydroxytryptamine, or serotonin, a neurotransmitter. Tryptophan must be hydroxylated to 5-hydroxytryptophan, which is decarboxylated to 5-hydroxytryptamine. Serotonin is then deaminated and oxidized to 5-hydroxyindoleacetic acid (5-HIAA). The urinary 5-HIAA excretion is elevated in the carcinoid syndrome.

Branched-Chain Amino Acid Catabolism

The three branched-chain amino acids—isoleucine, leucine, and valine—are all essential. They are degraded by transamination to α-keto acids, which are oxidized and decarboxylated to acyl-CoA.

The inherited absence of α-keto acid decarboxylase causes these three α-keto acids to accumulate and spill into the urine. Their sweet smell conferred the name of "maple-syrup urine disease" to this ketoaciduria. The disorder is fatal unless the infants afflicted by it are fed artificial proteins that are low in branched-chain amino acids.

Valine and isoleucine catabolism eventually produces propionyl-CoA, a three-carbon, fatty-acyl-CoA. This is carboxylated to methylmalonyl-CoA. A mutase employing vitamin B_{12} then converts this four-carbon compound to succinyl-CoA.

Methionine and Cysteine Metabolism

Methionine, an essential amino acid, is the precursor of cysteine (see p. 136).

As shown in Figure 12-4, adenosine from ATP joins via the sulfur atom of methionine to generate *S*-adenosylmethionine. This important methyl donor is then able to liberate its sulfur-bound methyl group in reactions such as the methylation of phosphatidylethanolamine to phosphatidylcholine or the methylation of norepinephrine to epinephrine. The *S*-adenosylhomocysteine generated in these reactions then liberates its adenosine group to yield homocysteine, which differs from cysteine only in having an additional CH_2 group in its side chain. Cystathionine synthase then condenses homocysteine with serine to form cystathionine. (Although cystathionine contains two amino acids, it is not a dipeptide, because they are not joined by a peptide linkage.) Cystathionase next cleaves α-ketobutyrate (originally from homocysteine) from the sulfur atom, leaving this sulfur attached to the former serine molecule, which now becomes cysteine.

Homocystinuria, which results from the absence of cystathionine synthase, and cystathioninuria, which results from the deficiency of cystathionase, both block cysteine synthesis, thus making it an essential amino acid in the strict sense. Both disorders respond to dietary methionine restriction and cysteine supplementation. Heterozygous homocystinuria is a common cause of atherosclerosis prior to age 50 among nonsmokers free of diabetes or hyperlipidemia.

In further catabolism, cysteine degradation liberates sulfate.

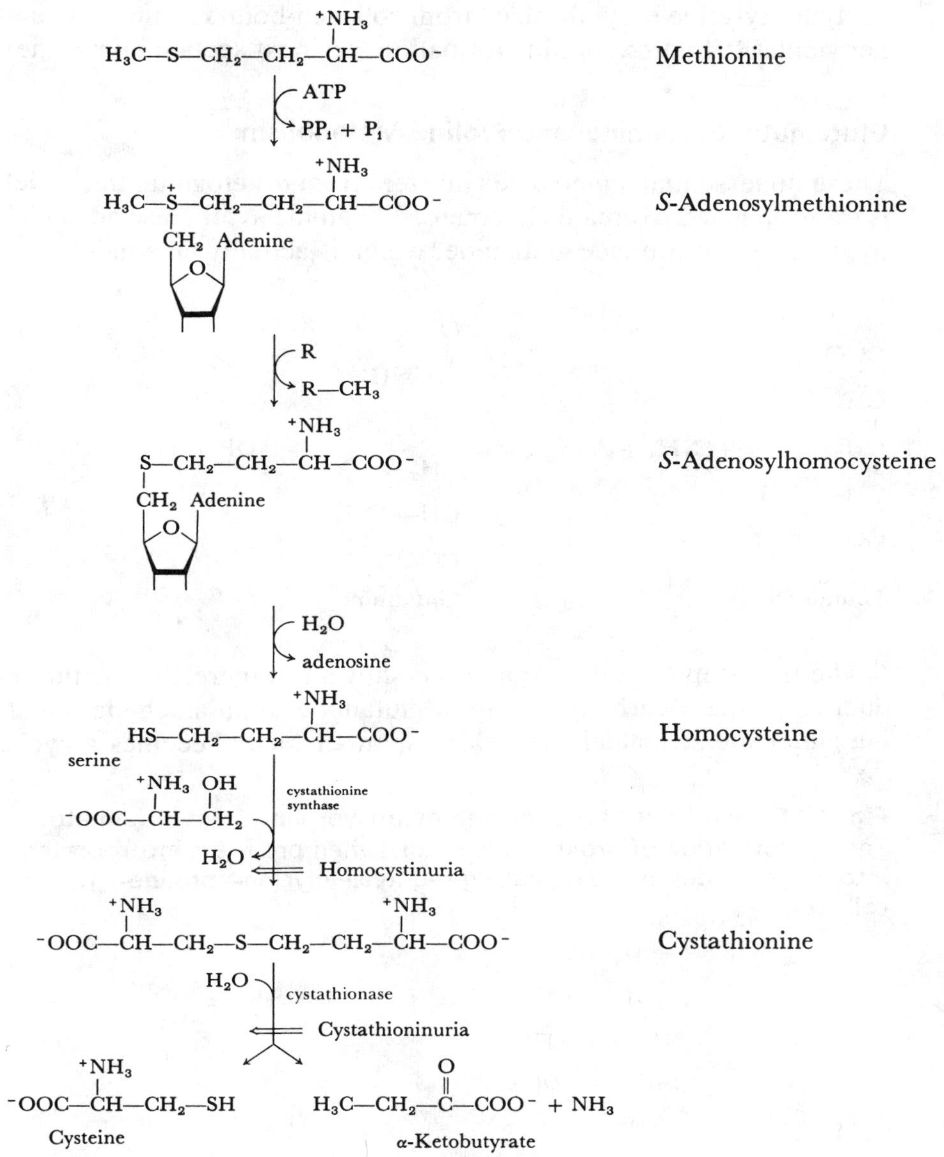

$$H_3C-S-CH_2-CH_2-\overset{\overset{+NH_3}{|}}{CH}-COO^-$$ Methionine

ATP

$PP_i + P_i$

$$H_3C-\overset{+}{S}-CH_2-CH_2-\overset{\overset{+NH_3}{|}}{CH}-COO^-$$ S-Adenosylmethionine

$\overset{|}{CH_2}$ Adenine

O

R

R—CH$_3$

$$S-CH_2-CH_2-\overset{\overset{+NH_3}{|}}{CH}-COO^-$$ S-Adenosylhomocysteine

$\overset{|}{CH_2}$ Adenine

O

H_2O

adenosine

$$HS-CH_2-CH_2-\overset{\overset{+NH_3}{|}}{CH}-COO^-$$ Homocysteine

serine

$$^-OOC-\overset{\overset{+NH_3}{|}}{CH}-\overset{\overset{OH}{|}}{CH_2}$$ cystathionine synthase

H_2O Homocystinuria

$$^-OOC-\overset{\overset{+NH_3}{|}}{CH}-CH_2-S-CH_2-CH_2-\overset{\overset{+NH_3}{|}}{CH}-COO^-$$ Cystathionine

H_2O cystathionase

Cystathioninuria

$$^-OOC-\overset{\overset{+NH_3}{|}}{CH}-CH_2-SH \qquad H_3C-CH_2-\overset{\overset{O}{\|}}{C}-COO^- + NH_3$$

Cysteine   α-Ketobutyrate

Fig. 12-4 Conversion of methionine to cysteine.

Basic Amino Acid Metabolism

The three basic amino acids—arginine, histidine, and lysine—are all essential. The urea cycle supplies only part of the daily arginine requirement.

Histidine decarboxylase transforms histidine into histamine, a mediator of the inflammatory response in allergic reactions.

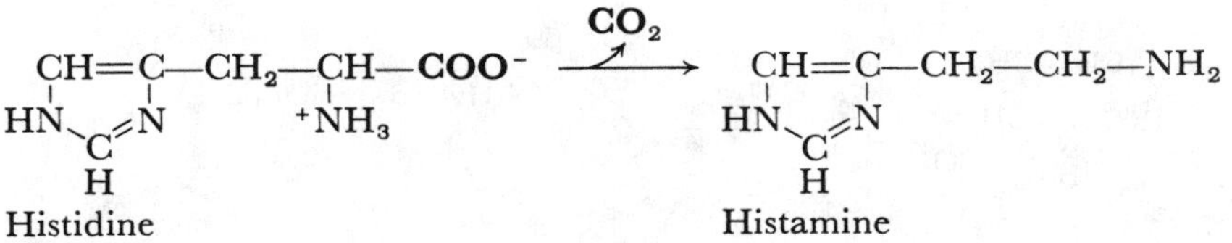

$$CH=C-CH_2-\overset{\overset{+NH_3}{|}}{CH}-COO^- \xrightarrow{\quad CO_2 \quad} CH=C-CH_2-CH_2-NH_2$$

HN   N   HN   N

C   C

H   H

Histidine   Histamine

Hydroxylysine is synthesized from collagen-bound lysine in a manner similar to hydroxyproline formation (see next section on proline).

Glutamate, Glutamine, and Proline Metabolism

These nonessential amino acids all stem from α-ketoglutarate, which is transaminated to create glutamate. Glutamine synthetase adds NH_3 to glutamate to produce glutamine, which is actually an amide:

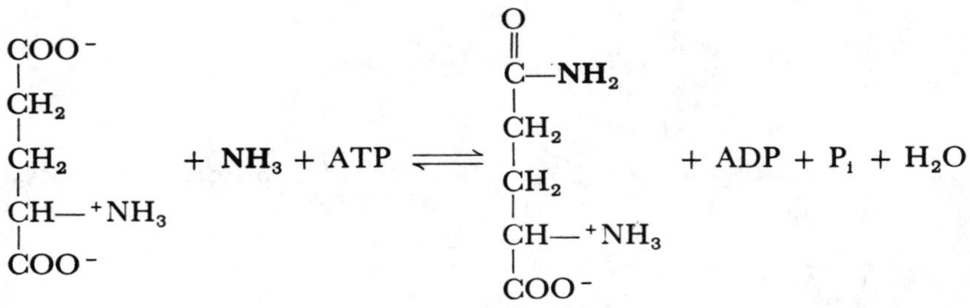

$$
\begin{array}{l}
\text{COO}^- \\
|\\
\text{CH}_2 \\
|\\
\text{CH}_2 \\
|\\
\text{CH}-{}^+\text{NH}_3 \\
|\\
\text{COO}^-
\end{array}
\quad + \text{NH}_3 + \text{ATP} \;\rightleftharpoons\;
\begin{array}{l}
\overset{\displaystyle O}{\overset{\|}{\text{C}}}-\text{NH}_2 \\
|\\
\text{CH}_2 \\
|\\
\text{CH}_2 \\
|\\
\text{CH}-{}^+\text{NH}_3 \\
|\\
\text{COO}^-
\end{array}
\quad + \text{ADP} + \text{P}_i + \text{H}_2\text{O}
$$

Glutamate   Glutamine

The first step in proline synthesis, shown in Figure 12-5, is the reduction of the γ-carboxyl group of glutamate to an aldehyde, creating glutamate semialdehyde, which spontaneously becomes a cyclic

Fig. 12-5 Synthesis of proline and hydroxyproline from glutamate. The hydroxylation of proline only occurs when proline is incorporated into polypeptides in the repeating sequence glycine—proline—Y, as in collagen.

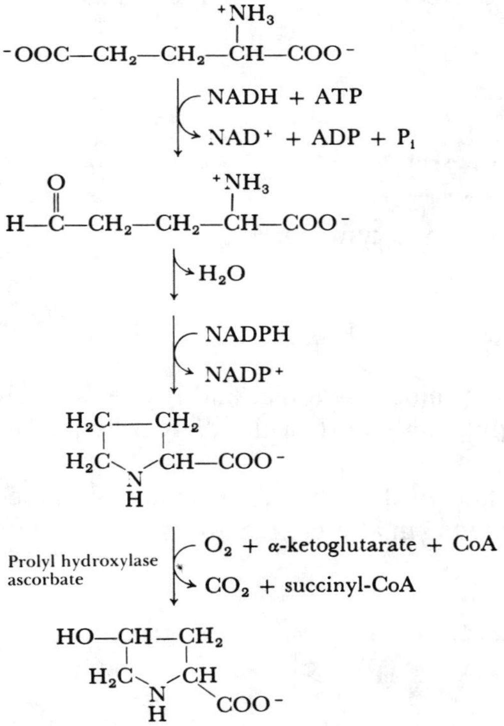

compound. Further reduction then yields the imino acid, proline. This pathway is reversible, allowing proline to be converted back to glutamate.

Only after incorporation into proteins such as collagen can proline be oxygenated to hydroxyproline. This reaction, shown in Figure 12-5, requires prolyl hydroxylase, O_2, and α-ketoglutarate, with iron and a reducing agent such as ascorbate as cofactors.

Glutamate, glutamine, and proline are all degraded to α-ketoglutarate, a TCA-cycle intermediate; hence they are glucogenic.

Aspartate, Asparagine, and Alanine Metabolism

Aspartate and alanine arise from the transaminations of oxaloacetate and pyruvate, respectively:

$$CH_3-\overset{\overset{\textstyle O}{\|}}{C}-COO^- \xrightarrow[\alpha\text{-ketoglutarate}]{\text{glutamate}} CH_3-\overset{\overset{\textstyle +NH_3}{|}}{C}H-COO^-$$

Pyruvate   Alanine

$$^-OOC-CH_2-\overset{\overset{\textstyle O}{\|}}{C}-COO^- \xrightarrow[\alpha\text{-ketoglutarate}]{\text{glutamate}} {}^-OOC-CH_2-\overset{\overset{\textstyle +NH_3}{|}}{C}H-COO^-$$

Oxaloacetate   Aspartate

Asparagine synthetase aminates aspartate to produce asparagine, the β-amide of aspartic acid. Conversely, asparaginase deaminates asparagine to aspartate.

Asparaginase is injected into children with acute lymphocytic leukemia to induce remission of the disorder by depriving leukemic cells of asparagine; normal cells are less sensitive to asparagine deprivation.

Glycine, Serine, and Threonine Metabolism

3-Phosphoglycerate from glycolysis is the precursor to serine.

Tetrahydrofolic acid (THFA) interconverts serine and glycine by removal or addition of a methylene group from N^5,N^{10}-methylene-THFA:

$$\overset{\overset{\textstyle OH}{|}}{C}H_2-\overset{\overset{\textstyle +NH_3}{|}}{C}H-COO^- + THFA \rightleftharpoons \overset{\overset{\textstyle +NH_3}{|}}{C}H_2-COO^- + CH_2\text{-}THFA + H_2O$$

Serine   Glycine

Threonine, an essential amino acid, is degraded to yield glycine and acetyl-CoA. Since glycine can be converted to 3-phosphoglycerate, threonine is glucogenic. The acetyl-CoA produced makes threonine also ketogenic, since this acetyl-CoA can be converted to ketone bodies.

Hormonal Control of Amino Acid Metabolism

Insulin and the somatomedins promote the active transport of amino acids across cell membranes. Insulin inhibits gluconeogenesis from amino acids, and it promotes protein synthesis. Growth hormone (STH), somatomedins, androgens, and thyroid hormones also promote a positive nitrogen balance by stimulating protein synthesis.

Glucocorticoids, on the other hand, enhance gluconeogenesis from amino acids. The resultant negative nitrogen balance in high-dose chronic glucocorticoid therapy leads to thinning of the skin and osteoporosis.

Following a meal, insulin release fosters tissue uptake of amino acids and subsequent protein synthesis. During fasting, however, the glucocorticoids predominate, and insulin virtually disappears. The protein in skeletal muscle and, eventually, in the viscera is eroded daily by gluconeogenesis to provide glucose. Humans have no amino acid storage capacity as such; useful tissue proteins must be sacrificed to supply glucose for the brain.

Dietary carbohydrate is protein-sparing. The presence of at least 100 g of dietary carbohydrate daily for adults prevents excess protein catabolism via gluconeogenesis. Much of the protein in a carbohydrate-deficient diet is not used for protein synthesis; instead, it enters gluconeogenesis to compensate for a lack of glucose.

Problems

Problem 1
Which class of amino acids contains only nonessential amino acids?

    A. Aromatic.
    B. Branched-chain.
    C. Basic.
    D. Sulfur-containing.
    E. Acidic.

Problems 2–6
Match the conversions below to the descriptions in Problems 2–6.

    A. Phenylalanine ⟶ phenylpyruvate
    B. Phenylalanine ⟶ tyrosine
    C. Dopa ⟶ dopamine
    D. Norepinephrine ⟶ epinephrine

2. The principal route of normal phenylalanine metabolism.
3. The alternate pathway of phenylalanine degradation in phenylketonuria.

4. Transamination.
5. Decarboxylation.
6. Methylation.

Problems 7–10
Match the compounds below to the disorders leading to their accumulation.

    A. Homogentisate
    B. Methylmalonic acid
    C. Branched-chain α-keto acids
    D. Homocysteine

  7. Maple-syrup urine disease.
  8. Homocystinuria.
  9. Alkaptonuria.
10. Vitamin B_{12} deficiency.

Problem 11
Histidine is converted to histamine by:

    A. Transamination.
    B. Hydroxylation.
    C. Decarboxylation.
    D. Reduction with NADH.
    E. Deamination.

Problem 12
The synthesis of hydroxyproline from proline in collagen requires all of the following except:

    A. Tetrahydrobiopterin.
    B. O_2.
    C. Reducing agents such as ascorbate.
    D. Iron.
    E. α-Ketoglutarate.

Problem 13
The methylene group transferred to glycine in converting it to serine comes from:

    A. *S*-Adenosylmethionine.
    B. N^5,N^{10}-Methylene-THFA.
    C. Methylene-B_{12}.
    D. Carboxybiotin.
    E. CO_2.

Problem 14
Choose the incorrect statement about carbamoyl-phosphate synthetase within the mitochondria.

    A. Produces carbamoyl phosphate for pyrimidine biosynthesis.
    B. Uses NH_4^+.

    C. Consumes 2 moles ATP per carbamoyl phosphate produced.
    D. Uses CO_2.
    E. Uses ATP.

Problem 15

A high intake of which amino acid can prevent pellagra in people consuming a niacin-deficient diet?

    A. Lysine.
    B. Methionine.
    C. Threonine.
    D. Tryptophan.
    E. Histidine.

Problem 16

Which of the following does not take part in the human urea cycle?

    A. Arginase.
    B. Aspartate.
    C. Argininosuccinate.
    D. Ornithine transcarbamoylase.
    E. Urease.

Answers

  1. E.
  2. B.
  3. A.
  4. A.
  5. C.
  6. D.
  7. C.
  8. D.
  9. A.
10. B.
11. C.
12. A. Tetrahydrobiopterin is used to hydroxylate phenylalanine, not proline.
13. B. When in doubt about a one-carbon transfer not involving CO_2, choose folic acid as the carrier. Vitamin B_{12} is known to be involved in only two reactions in humans. S-Adenosylmethionine is a methyl donor in a variety of reactions but not this one. Carboxybiotin transfers CO_2, not methylene groups.
14. A. Carbamoyl phosphate synthetase in the cytoplasm produces carbamoyl phosphate for pyrimidine synthesis.
15. D.
16. E. Urease in lower vertebrates and bacteria splits urea.

References

Devlin, T. M. *Textbook of Biochemistry with Clinical Correlations* (3rd ed.). New York: Wiley-Liss, 1992. Pp. 475–525.

Mathews, C. K., and van Holde, K. E. *Biochemistry.* Redwood City, Calif.: Benjamin/Cummings, 1990. Pp. 670–739.

Murray, R. K., Granner, D. K., Mayes, P. A., and Rodwell, V. W. *Harper's Biochemistry* (22nd ed.). Norwalk, Conn.: Appleton & Lange, 1990. Pp. 267–317.

Stryer, L. *Biochemstry* (3rd ed.). New York: Freeman, 1988. Pp. 495–515, 575–600.

Structure and Properties of Nucleic Acids

The fundamental components of nucleic acids are the pyrimidine and purine bases, the pentose sugars ribose and 2-deoxyribose, and phosphoric acid.

Pyrimidines

The **pyrimidine bases** contain the six-membered ring, with two nitrogen atoms, that constitutes the compound, pyrimidine. The three major pyrimidines found in humans—cytosine, uracil, and thymine—have oxygenated pyrimidine rings. Thymine, not to be confused with thiamine or vitamin B_1, has a methyl group at position 5. Cytosine has an amino group at position 4.

Oxygenated pyrimidines and purines exist in two different tautomeric forms, differing only with respect to the location of a proton. **Tautomers** are isomers that are freely interconvertible and exist in dynamic equilibrium under normal conditions. The keto or lactam ($NHC{=}O$) tautomer of cytosine is generally more prevalent than the enol or lactim ($N{=}COH$) tautomer at neutral pH.

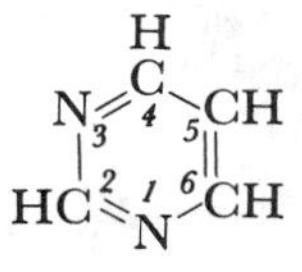

Pyrimidine

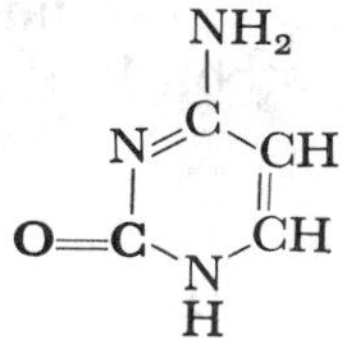

Cytosine
(lactam tautomer)

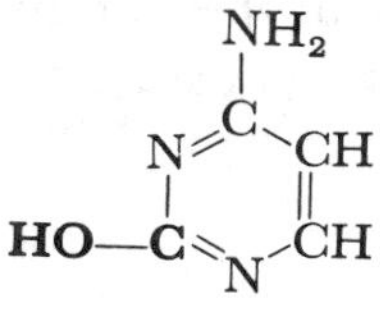

Cytosine
(lactim tautomer)

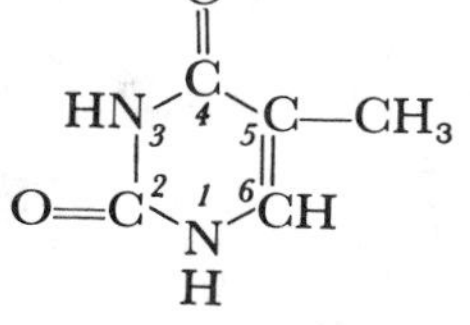

Thymine

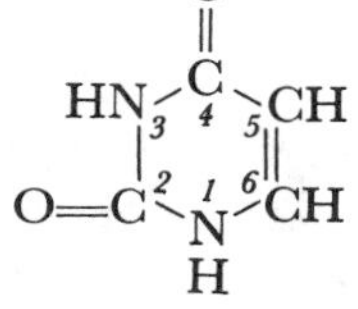

Uracil

Purines

The **purine bases** consist of a pyrimidine ring fused to an imidazole ring (histidine also has an imidazole ring). The two major purines in human nucleic acids are adenine and guanine. Adenine lacks a hydroxyl or a keto group and therefore cannot exhibit keto-enol tautomerism; it has an amino group at position 6. Guanine has a keto group at position 6 and an amino group at position 2.

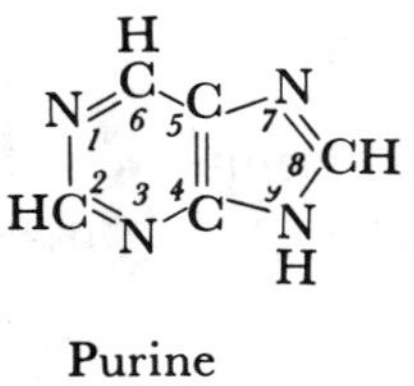

Purine 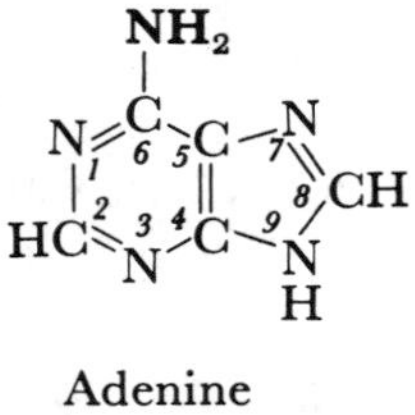   Adenine 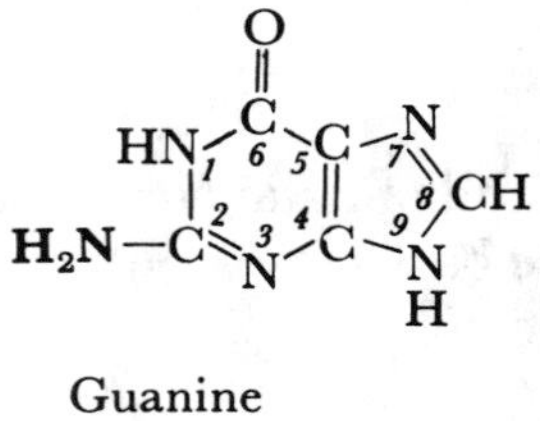   Guanine

Inosine, hypoxanthine, xanthine, and uric acid are purines created during the degradation of adenine and guanine.

The stimulant drugs caffeine (in coffee and tea) and theobromine (in chocolate) are methylxanthine compounds.

Nucleosides

A **nucleoside** consists of a purine or pyrimidine base bound to a pentose sugar. The nucleosides of ribose with adenine, guanine, cytosine, thymine, and uracil are called adenosine, guanosine, cytidine, thymidine, and uridine, respectively. If deoxyribose is present rather than ribose, the prefix **deoxy** is used, as in deoxyuridine (dU).

In numbering the atoms in nucleosides, a superscript prime is used to denote the atoms in the pentose. Thus, in cytidine, atom 1 of cytosine is joined by a glycosidic bond to carbon 1' of ribose.

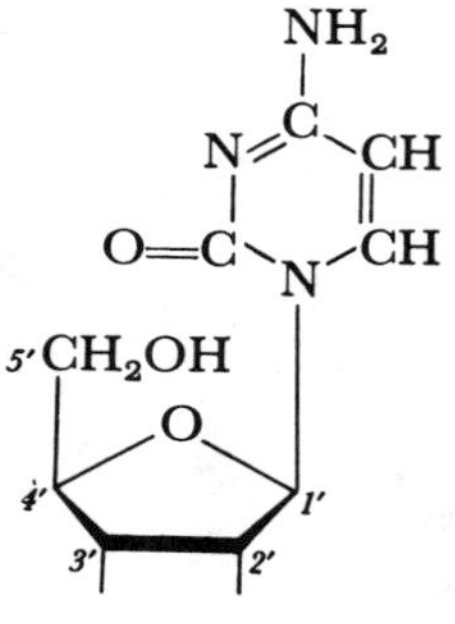

Cytidine 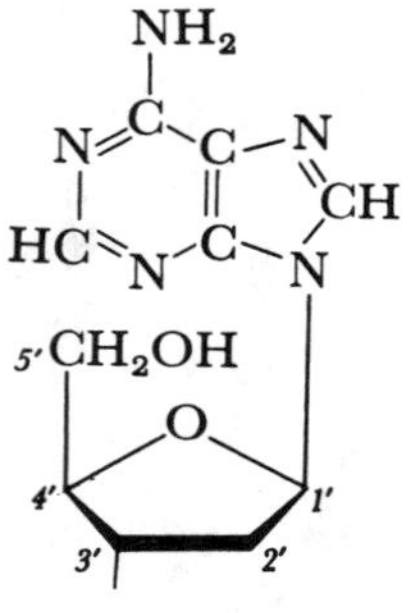   2'-Deoxyadenosine

Nucleotides

The terms "nucleotide" and "nucleoside phosphate" can be used interchangeably. In a **nucleotide,** a phosphate group is bound to the pentose sugar of a nucleoside. This phosphate group makes nucleotides strongly acidic.

The 5'-nucleoside monophosphate (NMP) of adenosine is called adenylic acid or adenosine monophosphate (AMP). Guanosine monophosphate (GMP) is called guanylic acid, cytidylic acid is CMP, uridylic acid is UMP, and thymidylic acid is TMP.

The 5'-nucleoside diphosphates (NDPs) are ADP, GDP, CDP, UDP, and TDP.

The 5'-nucleoside triphosphates (NTPs) are ATP, GTP, CTP, UTP, and TTP.

In 3',5'-cyclic AMP, the phosphate group joins to both the 3' and 5' carbons of ribose. This compound acts as a "second messenger" for many polypeptide hormones. For example, glucagon, a polypeptide hormone and first messenger, stimulates G proteins, which in turn activate adenyl cyclase in liver cell membranes to cleave ATP to 3',5'-cyclic AMP. This cyclic nucleotide in turn acts as a second messenger

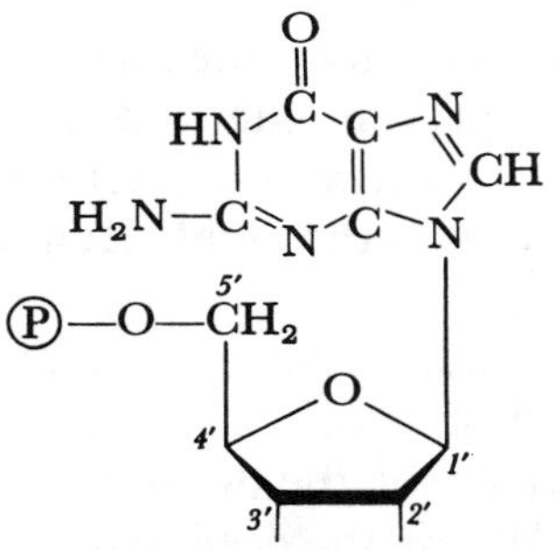

Guanylic acid (GMP) 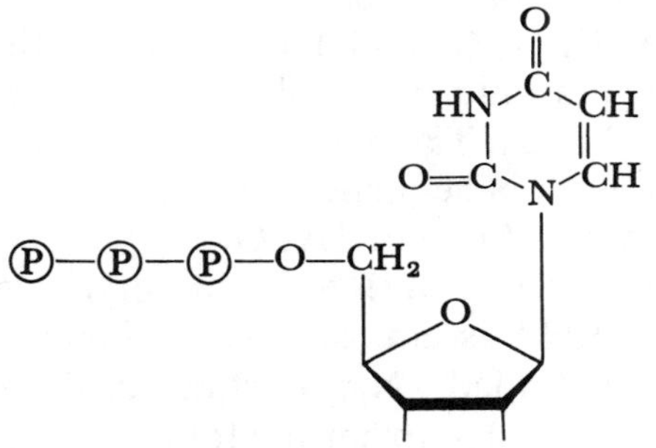   UTP

to activate glycogen phosphorylase and inhibit glycogen synthase, thereby freeing glucose so that its level in the blood may be maintained.

Polynucleotides and the Nucleic Acids

The nucleotides of a polynucleotide chain are linked to one another in 3',5'-phosphodiester bonds; phosphoric acid forms a phosphate ester to connect the 3' hydroxyl group of one pentose to the 5' carbon on another pentose.

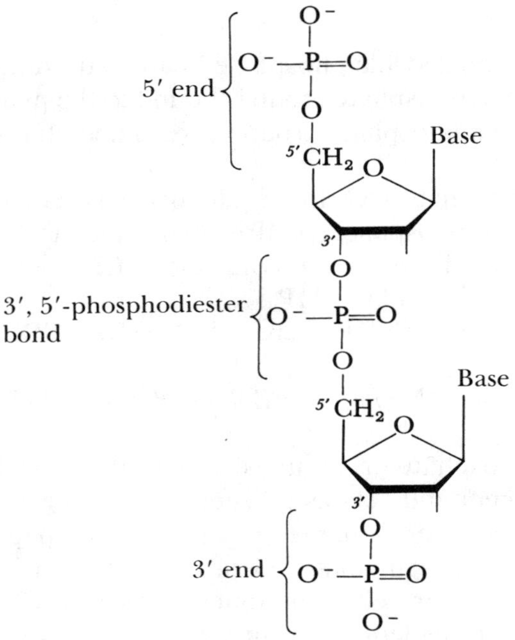

Structure of a dinucleotide

Polynucleotides can be represented diagrammatically using the letters A, C, G, U, and T to represent the bases. Vertical lines denote the sugars, and diagonal lines containing P represent the phosphodiester bonds. The 5' end of the polynucleotide is shown on diagrams and written in sequences to the left; the 3' end is to the right. Hence T is on the 5' end of TACG.

Ribonucleic acid (RNA) is a polynucleotide whose pentose sugar is ribose. RNA is almost always single-stranded; certain viruses have double-stranded RNA. Portions of RNA may assume the double helical structure due to hairpin folds in the chain. Within the double helix, guanine pairs with cytosine and adenine with uracil.

There are three basic types of RNA: messenger RNA (mRNA), ribosomal RNA (rRNA), and transfer RNA (tRNA). In addition to these three types, human cells also possess small RNA molecules.

Deoxyribonucleic acid (DNA) is a double-stranded polynucleotide with deoxyribose as its pentose.

The base content of DNA displays three sets of equivalent pairs:

1. A + G = T + C (purine content = pyrimidine content)
2. A = T
3. G = C

These equivalences suggested that the adenine of one DNA strand is always bound to thymine in the other strand and, similarly, that guanine is bound to cytosine. The proof of this base-pairing came when Watson and Crick proved by x-ray diffraction that the DNA structure was a double helix whose chains were complementary and antiparallel. By **complementary,** they meant that A binds to T and C to G between the chains. On the molecular level, this complementarity is accomplished by hydrogen bonding, as shown below. The chains are **antiparallel** because each end of the helix contains the 5′ end of one strand and the 3′ end of the other; hence, the chains travel in opposite directions.

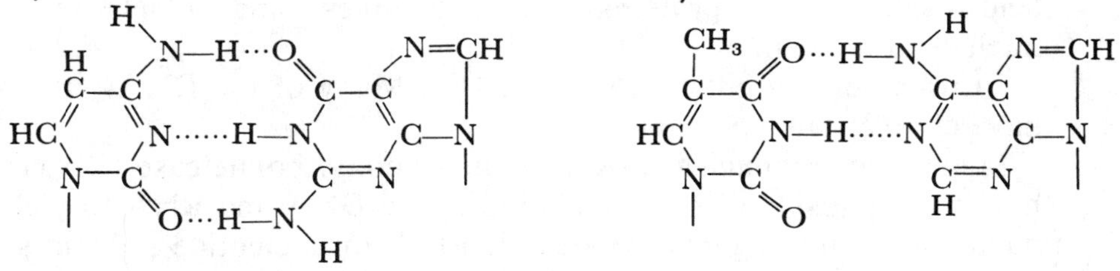

Hydrogen bonding of the base pairs of DNA

```
5'    d A—G—T—C—C    3'
          ⋮ ⋮ ⋮ ⋮ ⋮
3'    d T—C—A—G—G    5'
```
Antiparallel structure of DNA

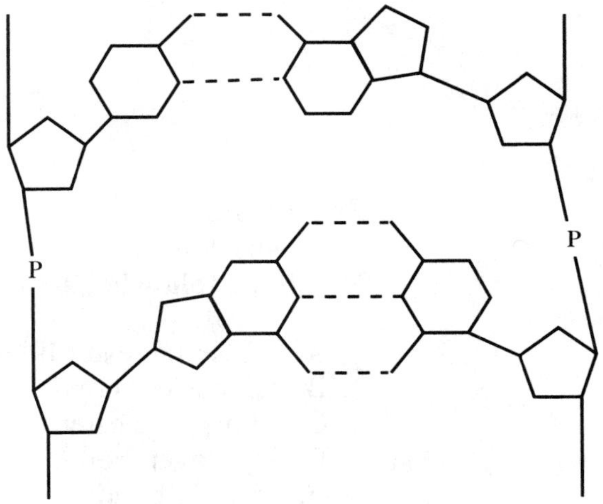

Hydrogen bonding of DNA base pairs

The two DNA chains form a helix about a common axis. The bases lie on the inside and the sugars on the outside of this double helix. The entire DNA molecule may be further coiled into a superhelix. Circular DNA refers to a continuous DNA chain; it may take any conformation.

The most common form for DNA is the double helix described by Watson and Crick, known as B-DNA. B-DNA can be bent to form circular DNA or it can be supercoiled. It can also be kinked by certain base sequences such as four or more consecutive adenine residues. B-DNA contains a regular series of grooves. Each adenine-thymine base pair forms a major groove, whereas each cytosine-guanine base pair forms a minor groove.

Less common than B-DNA is A-DNA and Z-DNA. A-DNA differs from B-DNA in that its right-handed double helix is shorter but broader and the base pairs are tilted. A-DNA occurs when DNA is dehydrated. DNA-RNA hybrids and double-stranded RNA have a double helical structure similar to A-DNA.

Z-DNA is a left-handed double helix, in contrast to the right-handed double helix of A- and B-DNA. Z-DNA forms when the base sequence alternates regularly from a purine to a pyrimidine.

If hydrogen bonds between base pairs are broken, the double helix separates into two polynucleotides. As temperature rises this process of melting occurs suddenly.

RNA and DNA are strongly acidic due to their many phosphoric acid residues. Basic proteins, such as histones, bind to DNA in the nucleus of the cell.

Viruses contain either RNA or DNA. Many of the DNA viruses have circular DNA.

Pancreatic ribonuclease (RNase) and deoxyribonuclease (DNase) hydrolyze phosphodiester bonds in dietary DNA and RNA to yield nucleotides and oligonucleotides (short-chain nucleotide polymers). Phosphodiesterase, found in various human tissues, also hydrolyzes phosphodiester bonds.

Problems

Problems 1–4

Match the following items to their descriptions in Problems 1–4:

  A. Phosphodiester bond
  B. Glycosidic bond
  C. Phosphate ester
  D. Hydrogen bonds
  E. Peptide bonds

1. Links nucleoside to phosphoric acid.
2. Holds DNA strands together.
3. Joins the base to the pentose sugar.
4. Links nucleosides in RNA to one another.

Problem 5

DNA:

  A. Contains equivalent amounts of adenine and guanine.
  B. Melting is due to cleavage of phosphodiester bonds.
  C. Strands are complementary and parallel to one another.
  D. Unlike RNA is not cleaved by phosphodiesterase.
  E. Strands are always joined by hydrogen bonds.

Problems 6–8

Match the structures below to their descriptions given in Problems 6–8:

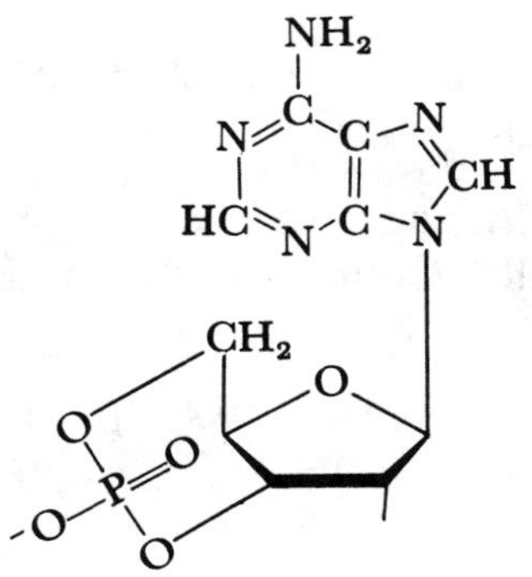

A.

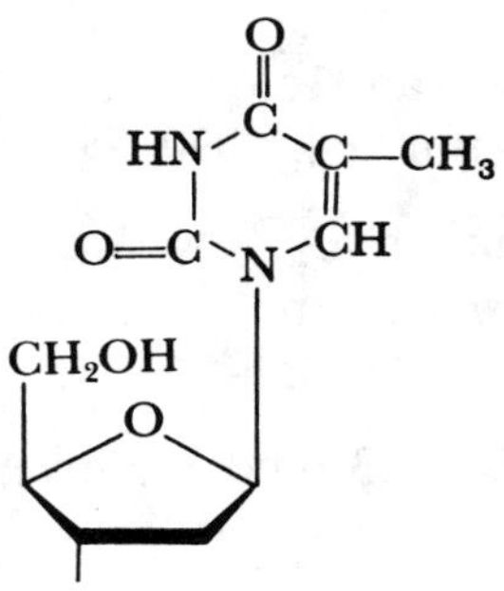

B.

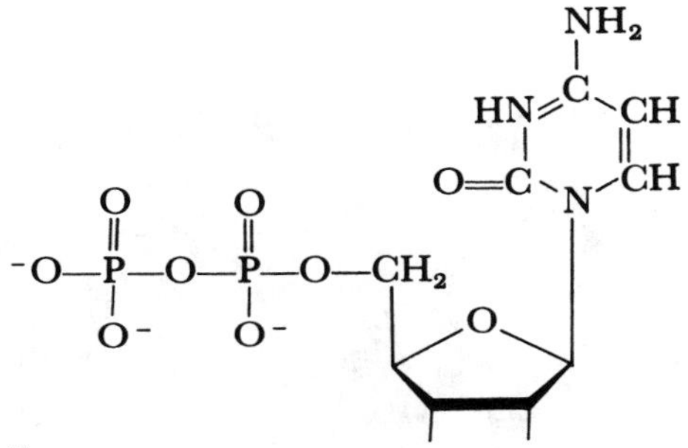

C.

6. A cytidine nucleotide.
7. 3',5'-Cyclic AMP.
8. Thymidine.

Answers

1. C.
2. D.
3. B.
4. A.
5. E. The amount of adenine plus guanine equals that of cytosine plus thymine. Cleavage of hydrogen bonds between base pairs is responsible for melting. The strands run antiparallel to one another. The phosphodiester linkages of both DNA and RNA are cleaved by phosphodiesterase.
6. C. This is CTP.
7. A.
8. B.

References

Devlin, T. M. *Textbook of Biochemistry with Clinical Correlations* (3rd ed.). New York: Wiley-Liss, 1992. Pp. 607–648, 682–692, 1157–1158.

Mathews, C. K., and van Holde, K. E. *Biochemistry.* Redwood City, Calif.: Benjamin/Cummings, 1990. Pp. 91–120.

Murray, R. K., Granner, D. K., Mayes, P. A., and Rodwell, V. W. *Harper's Biochemistry* (22nd ed.). Norwalk, Conn.: Appleton & Lange, 1990. Pp. 333–341, 356–365.

Stryer, L. *Biochemistry* (3rd ed.). New York: Freeman, 1988. Pp. 71–90, 650–655.

Nucleoside Catabolism and Biosynthesis

14

The digestion of dietary nucleoproteins begins with their proteolysis in the stomach and intestine. Once liberated from their protein coating, the polynucleotides are cleaved by pancreatic ribonucleases and deoxyribonucleases to oligo- and mononucleotides. Mononucleotidases in the intestine cleave mononucleotides to mononucleosides plus free phosphate. Intestinal nucleosidases then break the nucleosides into free bases plus pentose sugars. After absorption, most dietary nucleotides are catabolized rather than incorporated into human nucleic acids. Humans do not require any dietary purines or pyrimidines.

Pyrimidine Biosynthesis

Required in both pyrimidine and purine biosynthesis is 5-phosphoribosylpyrophosphate (PRPP), which is generated by pyrophosphorylating ribose-5-P with ATP. The hexose-monophosphate shunt provides ribose-5-P. The enzyme that catalyzes this pyrophosphorylation, ribose-phosphate pyrophosphokinase, is allosterically inhibited by ADP and GDP.

$$\text{Ribose-5-P} + \text{ATP} \longrightarrow \text{PRPP} + \text{AMP}$$

Pyrimidine synthesis also requires carbamoyl phosphate. Carbamoyl-phosphate synthetase (glutamine) in the cytoplasm deaminates glutamine and combines the NH_3 obtained with CO_2 and H_2O to yield carbamoyl phosphate. The urea cycle in the mitochondria utilizes another enzyme, carbamoyl-phosphate synthase (ammonia), which does not require glutamine (see Fig. 12-2). This difference between these two enzymes is indicated by appending the name of the substrate in parentheses.

$$\text{Glutamine} + CO_2 + H_2O + 2\text{ATP} \longrightarrow \text{glutamate} + \text{carbamoyl phosphate} + 2\text{ADP} + P_i$$

Aspartate transcarbamoylase, the rate-controlling enzyme in pyrimidine synthesis, links the carbamoyl group of the unstable carbamoyl phosphate to aspartate to generate *N*-carbamoylaspartate, as shown in Figure 14-1. In bacteria but not mammals, CTP is an inhibitory modifier for this allosteric enzyme. Since *N*-carbamoylaspartate is used only in pyrimidine biosynthesis, this is the committed step of the pathway.

Dehydration transforms the straight-chain *N*-carbamoylaspartate into a cyclic pyrimidine, dihydroorotic acid, which is oxidized with NAD^+ to orotic acid. Next, orotate phosphoribosyltransferase adds PRPP to orotic acid to produce orotidine-5-P, a nucleoside phosphate or nucleotide. Orotidine-5-P decarboxylase then removes the carboxyl group from the side chain to create uridylic acid or UMP.

The rare, hereditary absence of orotate phosphoribosyltransferase and orotidine-5-P decarboxylase blocks UMP synthesis and leads to orotic aciduria. One protein serves as both of these enzymes. This disease is treated by administering large doses of uridine to bypass this enzymatic block and supply a precursor for synthesizing UTP,

Fig. 14-1 Pyrimidine biosynthesis.

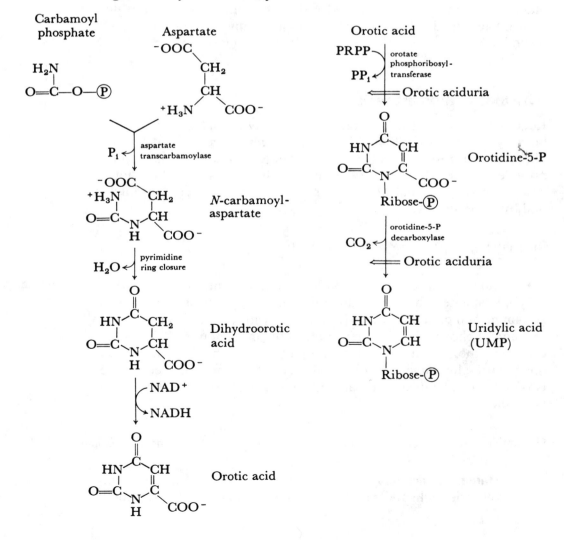

CTP, and TTP. This CTP will then allosterically inhibit aspartate transcarbamoylase, thereby reducing orotic-acid production.

UMP is the precursor of all the pyrimidine nucleotides. ATP phosphorylates UMP first to UDP and then to UTP:

$$\text{UMP} \xrightarrow[\text{ATP} \quad \text{ADP}]{} \text{UDP} \xrightarrow[\text{ATP} \quad \text{ADP}]{} \text{UTP}$$

The uracil of UTP is aminated with glutamine to create the cytosine in CTP:

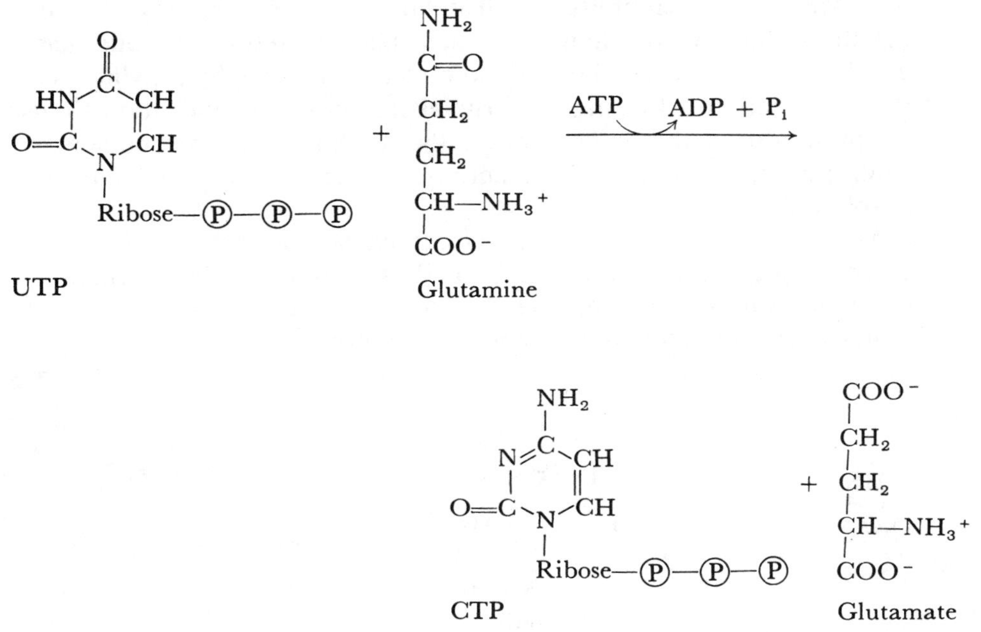

UTP   Glutamine

CTP   Glutamate

5-Iodo-2′-deoxyuridine is a pyrimidine analog used clinically to treat keratitis due to herpes virus. This iodinated pyrimidine becomes incorporated into viral DNA during DNA replication and interferes with DNA function.

Synthesis of Deoxyribonucleotides

In humans, all the nucleotides are synthesized initially as ribonucleotides. Once this synthesis is complete, the 2′ hydroxy group of ribose is reduced to a hydrogen atom. In the first step of this reduction, thioredoxin reductase accepts an electron from NADPH. The electron is next transferred to thioredoxin. In the final step, ribonucleotide reductase uses this electron to reduce a ribonucleotide to a deoxyribonucleotide.

Thymidylate synthetase uses N^5,N^{10}-methylene-THFA to methylate dUMP to create dTMP and dihydrofolic acid (DHFA):

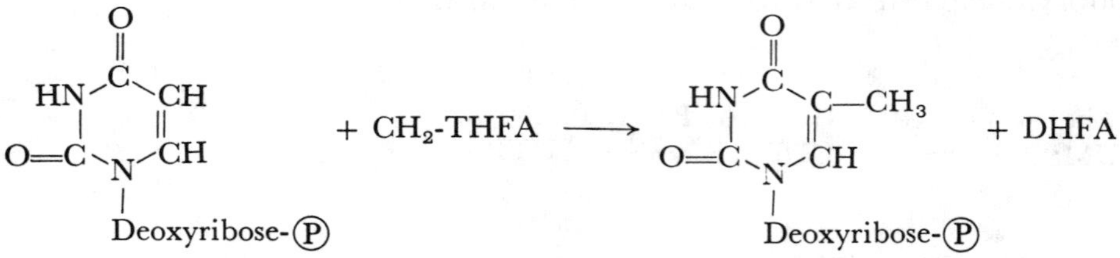

Folic acid antagonists, e.g., methotrexate, are used to treat leukemia. Methotrexate inhibits dihydrofolate reductase, thereby preventing the reduction of dihydrofolic acid back to tetrahydrofolic acid. The lack of tetrahydrofolic acid indirectly blocks the thymidylate synthetase reaction. Blocking this crucial step in DNA replication slows the proliferation of both leukemic cells and those of normal bone marrow; the latter effect causes macrocytic anemia similar to that seen in folate deficiency.

5-Fluorouracil (5-FU) is used to treat many solid tumors in humans. Thymidylate synthetase converts 5-FU to 5-fluoro-dUMP, which irreversibly inhibits the enzyme. Because 5-FU causes an enzyme to "kill" itself, it is known as a suicide inhibitor.

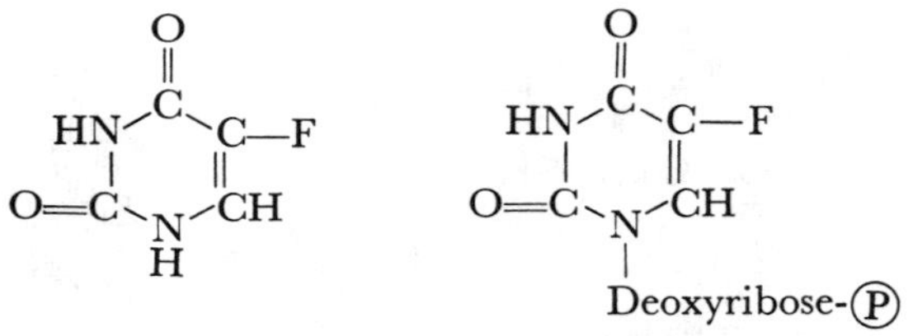

Cytosine arabinoside, like 5-fluorouracil, is a pyrimidine analog. Its medical use lies in treating acute myelogenous leukemia. Although its exact mechanism of action is uncertain, the arabinose sugar that replaces the normal deoxyribose interferes with base stacking when this analog is incorporated into DNA.

In pyrimidine catabolism, β-alanine arises from uracil and cytosine whereas β-aminoisobutyrate arises from thymine. The pyrimidine nitrogens are liberated as ammonia, which can enter the urea cycle.

Purine Synthesis

Since purines contain both a pyrimidine and an imidazole ring, their synthesis is more complex than that of pyrimidines. In purine synthesis, ribose is bound to the growing purine ring from the outset rather than being attached after ring formation, as occurs in pyrimidine synthesis. Thus, all the intermediates in purine synthesis contain ribose phosphate.

Biosynthesis of purines, like that of pyrimidines, requires PRPP. Unlike pyrimidine biosynthesis, tetrahydrofolic acid is required. The first purine produced is inosinic acid or IMP. AMP and GMP are then produced from IMP in separate pathways. ATP is used as a substrate to produce GMP; GTP is used as a substrate to produce AMP. Thus, a shortage of ATP will lead to a compensatory increase in AMP formation.

GMP provides negative feedback to regulate its own formation by inhibiting the oxidation of IMP to xanthylic acid. Similarly, AMP inhibits the formation of its precursor, adenylosuccinate, from IMP. Folic acid antagonists block the two folate-dependent steps in IMP synthesis.

Two purine analogs used clinically to treat leukemia are **6-mercaptopurine** and **thioguanine,** analogs of hypoxanthine and guanine, respectively. Both are activated in vivo after joining ribose phosphate. 6-Mercaptopurine bound to ribose phosphate inhibits the conversions of IMP to both xanthylic acid and adenylosuccinate, as well as inhibiting the amidophosphoribosyltransferase reaction.

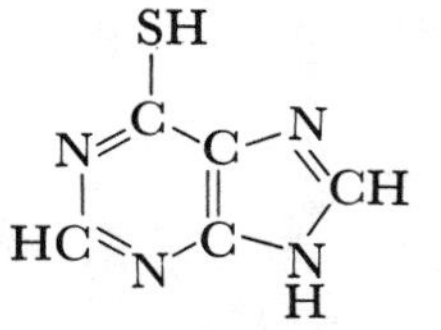

6-Mercaptopurine 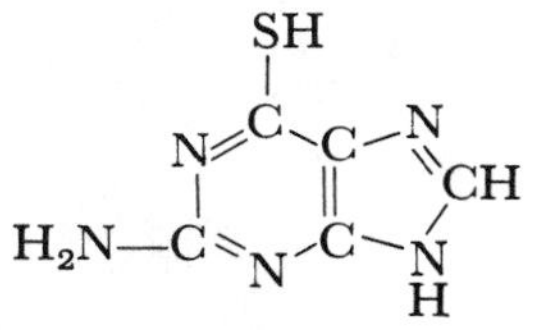   Thioguanine

Nucleosidemonophosphate kinase phosphorylates any nucleoside monophosphate to its diphosphate form using ATP. Similarly, nucleosidediphosphate kinase transforms nucleoside diphosphates to triphosphates:

$$NMP + ATP \xrightleftharpoons{\text{nucleosidemonophosphate kinase}} NDP + ADP$$

$$NDP + ATP \xrightleftharpoons{\text{nucleosidediphosphate kinase}} NTP + ADP$$

Purine Catabolism and Salvage

AMP and GMP released from nucleic acids by the action of nucleases are hydrolyzed to yield the nucleosides adenosine and guanosine, as shown in Figure 14-2. The high-energy phosphate groups expended in phosphorylating the nucleosides are not regained in the reverse reactions. These nucleosides then liberate their ribose, leaving adenine and guanine.

Two salvage pathways exist to regenerate AMP and GMP from adenine and guanine, respectively. **Adenine phosphoribosyltransferase**

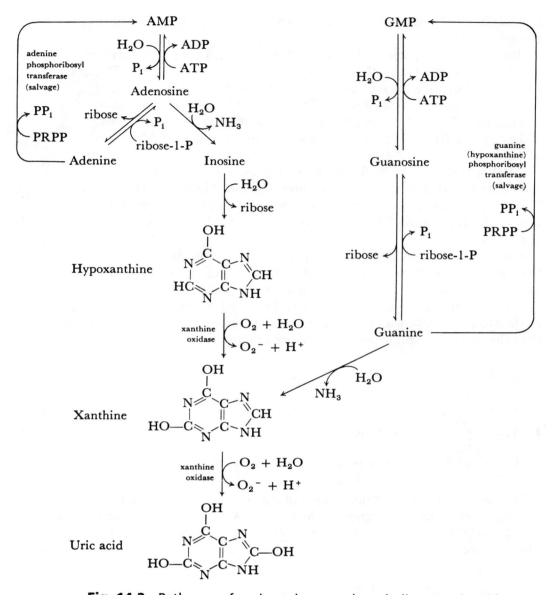

Fig. 14-2   Pathways of purine salvage and catabolism to uric acid.

adds ribose phosphate from PRPP to adenine to regenerate AMP. **Hypoxanthine-guanine phosphoribosyltransferase** catalyzes the analogous reaction with guanine or hypoxanthine.

$$\text{Adenine} + \text{PRPP} \xrightarrow{\text{Adenine phosphoribosyltransferase}} \text{AMP} + \text{PP}_i$$

$$\text{Guanine} + \text{PRPP} \xrightarrow{\text{Hypoxanthine-guanine phosphoribosyltransferase}} \text{GMP} + \text{PP}_i$$

In **Lesch-Nyhan syndrome,** an inherited absence of hypoxanthine-guanine phosphoribosyltransferase, excessive purine synthesis results in order to compensate for the lack of guanine salvage. The enhanced purine synthesis leads to increased purine catabolism and consequent uric-acid production. Hyperuricemia (high serum uric acid concentration) ensues, along with self-mutilation and mental retardation.

A second pathway for purine salvage adds ribose-1-P to adenine and guanine to create adenosine and guanosine, respectively, which are phosphorylated with ATP to yield AMP and GMP.

Besides being converted to adenine, adenosine may also be deaminated to inosine. Inosine then liberates its ribose, forming hypoxanthine. **Xanthine oxidase,** an enzyme that contains molybdenum and iron, adds O_2 and water to hypoxanthine to create xanthine and the superoxide radical O_2^-, which is converted to H_2O_2 by superoxide dismutase. Catalase then splits hydrogen peroxide to yield H_2O + $\frac{1}{2}O_2$. Guanine, too, may be deaminated to produce xanthine. In the final step of human purine catabolism, xanthine oxidase transforms xanthine into uric acid.

Once it is synthesized, humans cannot degrade the purine ring itself. Thus, the purine derivative uric acid is the endproduct of human purine catabolism. The human kidney disposes of most of man's uric acid by glomerular filtration and tubular secretion. The renal tubules, however, reabsorb much of the filtered uric acid.

Approximately one quarter of man's daily uric acid removal occurs via intestinal bacteria, which possess urate oxidase to transform uric acid into allantoin. Other bacterial enzymes can degrade allantoin to allantoic acid and then to urea. Certain bacteria can then split urea to ammonia and carbon dioxide.

Gout, caused by hyperuricemia, is among the most common metabolic disorders in America, affecting about 1 in 200 adults. The clinical expressions of gout stem from the low solubility of sodium urate, which precipitates in joints (gouty arthritis), connective tissue (tophi), and the kidneys (urate stones). Hyperuricemia may be due to the overproduction of purines or the underexcretion of urate by the kidneys, or both. The overactivity of two enzymes in purine synthesis can cause human gout: ribose-phosphate pyrophosphokinase, which synthesizes PRPP, and amidophosphoribosyltransferase, which catalyzes the rate-limiting step in IMP synthesis.

An excess supply of ribose-5-P occurs in glucose-6-phosphatase deficiency (type-I glycogen storage disease) because the accumulating glucose-6-P is converted to ribose-5-P by the pentose-phosphate pathway. This, too, causes hyperuricemia.

Defective guanine salvage in Lesch-Nyhan syndrome also causes purine overproduction, and excessive purine synthesis occurs in diseases involving rapid cellular proliferation, such as leukemia.

The renal underexcretion of urate occurs in many kidney diseases. Substances that inhibit the tubular secretion of urate include the thiazide and "loop" diuretics, β-hydroxybutyrate (which abounds in starvation and diabetic ketoacidosis), and lactic acid (which is found in high levels in lactic acidosis and lactic acidemia, as well as in glucose-6-phosphatase deficiency).

Allopurinol is an analog to hypoxanthine that is used clinically to reduce the serum uric-acid level. It is not a true purine, because it lacks an imidazole ring (note the transposition of the C and N atoms in the formulas below). Allopurinol is a suicide inhibitor of xanthine oxidase. It is converted by xanthine oxidase to alloxanthine, which binds to the active site of xanthine oxidase, thereby inhibiting the enzyme. After allopurinol is administered, the composition of the urinary purines excreted changes from predominantly uric acid to a mixture of hypoxanthine, xanthine, and uric acid. Being far more soluble than uric acid, hypoxanthine and xanthine do not readily precipitate to form kidney stones.

Allopurinol   Hypoxanthine

Problems

Problems 1–4

Match the enzymes below to the descriptions in Problems 1–4. Each enzyme may be used more than once or not at all.

    A. Orotidine-5-P decarboxylase
    B. Aspartate transcarbamoylase
    C. Carbamoyl-phosphate synthetase (glutamine)
    D. Carbamoyl-phosphate synthase (ammonia)
    E. Orotate phosphoribosyltransferase

1. Inherited dual enzyme deficiency causes inadequate pyrimidine synthesis (two answers).
2. Supplies nitrogen to the urea cycle.
3. Rate-controlling step of pyrimidine biosynthesis.
4. Generates the unstable carbamoyl phosphate for pyrimidine biosynthesis.

Problem 5

Which statement about aspartate transcarbamoylase is *incorrect*?

    A. It obeys Michaelis-Menten kinetics.
    B. Its allosteric inhibitor is CTP.
    C. It catalyzes the committed step of pyrimidine biosynthesis.
    D. It is a multiple-subunit enzyme.
    E. It synthesizes *N*-carbamoylaspartate.

Problem 6

Purine biosynthesis:

    A. Purines are assembled on ribose phosphate.
    B. Is blocked by 5-fluorouracil.
    C. Is reduced by a deficiency of hypoxanthine-guanine phosphoribosyltransferase.

D. Is not altered by thioguanine.

E. Includes the thymidylate synthesis reaction.

Problem 7

Pyrimidine analogs used to treat human diseases include all of the following except:

A. 3′-azido-3′-deoxythymidine (AZT) for acquired immunodeficiency syndrome (AIDS).

B. 5-fluorouracil for a variety of solid tumors.

C. Acyloguanosine (acyclovir) for herpes simplex virus.

D. Cytosine arabinoside for myeloid leukemia.

Answers

1. A, E. The disease is orotic aciduria.
2. D. This mitochondrial enzyme uses NH_3 as its nitrogen source; the carbamoyl phosphate produced enters the urea cycle.
3. B.
4. C. This cytoplasmic enzyme deaminates glutamine to supply NH_3 for carbamoyl-phosphate synthesis.
5. A. No allosteric enzyme obeys Michaelis-Menten kinetics.
6. A. 5-Fluorouracil blocks pyrimidine biosynthesis. A deficiency of hypoxanthine-guanine phosphoribosyltransferase enhances purine biosynthesis. Thioguanine blocks purine biosynthesis. Thymidylate synthesis creates dTMP, a pyrimidine nucleotide.
7. C. This is a purine analog.

References

Devlin, T. M. *Textbook of Biochemistry with Clinical Correlations* (3rd ed.). New York: Wiley-Liss, 1992. Pp. 529–571.

Mathews, C. K., and van Holde, K. E. *Biochemistry*. Redwood City, Calif.: Benjamin/Cummings, 1990. Pp. 742–776.

Murray, R. K., Granner, D. K., Mayes, P. A., and Rodwell, V. W. *Harper's Biochemistry* (22nd ed.). Norwalk, Conn.: Appleton & Lange, 1990. Pp. 342–355.

Stryer, L. *Biochemistry* (3rd ed.). New York: Freeman, 1988. Pp. 601–626.

DNA Replication and Transcription

DNA Replication

The double helix of DNA is replicated in a **semiconservative** manner; that is, each strand fathers a complementary strand so that the offspring DNA contain one parental strand and one newly synthesized strand. Human DNA, unlike that of *Escherichia coli,* is intertwined with basic proteins called **histones,** and it requires "unmasking" before DNA replication can begin.

To begin DNA replication, **unwinding proteins** open segments along the DNA double helix. Once part of the helix has been unwound, an **RNA polymerase** called **primase** creates a complementary RNA primer, of 5 to 12 bases for each DNA strand. Like all enzymes that synthesize polynucleotides, primase cleaves pyrophosphate from nucleoside triphosphates (NTP) as it adds nucleoside monophosphates (NMP) to lengthen the chain antiparallel to the DNA template (i.e., 3′ to 5′ along the DNA template, but 5′ to 3′ along the emerging RNA primer).

$$\text{NTP + Polynucleotide} \xrightarrow{\text{Primase}} \text{NMP—Polynucleotide + PP}_i$$

$$\text{PP}_i + \text{H}_2\text{O} \longrightarrow 2\,\text{P}_i$$

Once this temporary RNA primer has been synthesized, DNA polymerase can begin the real work of DNA replication. DNA is synthesized at the sites of unwinding, termed replication forks. In *E. coli* there are two replication forks for every DNA molecule, which move in opposite directions. DNA polymerase is located at the replication forks (Fig. 15-1).

DNA replication differs for the two parent DNA strands. On the leading strand, the 3′ end is open, allowing DNA replication by **DNA polymerase** to proceed continuously in a 5′ to 3′ direction starting at the end of the RNA primer. The addition of nucleotides is fueled by the hydrolysis of pyrophosphate liberated from the nucleotide triphosphates. *E. coli* has three forms of DNA polymerase. DNA polymerase III has the highest catalytic activity of these forms because it is highly **processive:** it remains bound to its substrate during many cycles of polymerization. DNA polymerase I is only slightly processive and is primarily involved in repairing DNA: it acts as an exonucleave to excise incorrect nucleotides from the 3' end of the primer. After the leading strand has been replicated the RNA primer is hydrolyzed.

On the lagging strand, however, only the 5' end is open. Since DNA polymerase cannot operate in a 3' to 5' direction, *the lagging strand must be synthesized discontinuously 3' to 5' in pieces known as* **Okazaki fragments.** Each fragment consists of 1000 to 2000 nucleotides in prokaryotes and 100–200 nucleotides in eukaryotes. Later DNA polymerase fills in the gaps between the Okazaki fragments with deoxyribonucleotides. Finally, **DNA ligase** joins the ends of these fragments to create an entire DNA chain.

Note that DNA polymerases cannot initiate DNA replication. Instead, primase begins this process by creating an RNA primer that is later excised.

Thus the major events in DNA replication occur in this sequence:

1. Unwinding proteins open the DNA double helix.
2. Primase creates a temporary RNA primer for each DNA strand.
3. DNA polymerase at the replication fork synthesizes DNA in a 5' to 3' direction. The leading strand can be synthesized continuously. The lagging strand, however, must be synthesized discontinuously in the form of Okazaki fragments.
4. DNA polymerase then removes the RNA primer and fills the gaps between Okazaki fragments.
5. DNA ligase then joins DNA fragments of the lagging strand, creating a single DNA molecule.

Newly replicated DNA is further modified by a class of enzymes known as **topoisomerases.** Type I topoisomerases bind to DNA, break one strand, and change the conformation of the molecule prior to restoring the chain. The net effect is to loosen the superhelical structure. Type II topoisomerases such as DNA gyrase break both DNA strands prior to rejoining the strands. DNA gyrase uses the energy liberated by ATP hydrolysis to add torsion in the form of negative supercoils to the resultant DNA.

Fig. 15-1   Schematic diagram of a replication fork. (From *Biochemistry,* 3rd ed. By Lubert Stryer. Copyright © 1988 by Lubert Stryer. Reprinted with permission of W. H. Freeman and Company.)

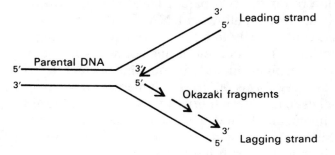

DNA Repair

Agents that damage DNA include ultraviolet radiation, x-ray irradiation, heat, extreme pH, and chemicals that alter purines and pyrimidines. One of the most frequent causes of DNA damage is ultraviolet radiation, which induces adjacent thymine rings to dimerize via covalent bonds. Thymine dimers block both DNA replication and transcription because they do not fit into the double helix. During the excision repair of DNA to remove thymine dimers, **excinuclease** cleaves the involved DNA strand at two sites, removing a segment with 12 nucleotides. DNA polymerase then synthesizes a new DNA segment to replace the one excised. Finally, DNA ligase joins the 3′ end of the new segment to the appropriate place on the unaltered DNA.

Critical to DNA repair is a mechanism for identifying which of the two DNA strands is damaged. Methylated adenine groups in the sequence GATC mark the original DNA strand because newly synthesized strands are not methylated. Hence enzymes repairing DNA seek out the unmethylated strand.

Xeroderma pigmentosum, a rare human disease leading to photosensitivity, stems from inherited excinuclease deficiency. Thymine dimers created after solar radiation cannot, therefore, be adequately repaired.

Cytosine in DNA is not stable; some is spontaneously converted to uracil. Uracil binds to adenine, whereas cytosine binds to guanine. Thus the degradation of cytosine to uracil would introduce mutations. A mechanism for replacing uracil in DNA with cytosine is therefore important. Uracil-DNA glycosidase removes uracil from DNA but not from RNA. DNA polymerase then replaces it with cytosine. Lastly DNA ligase closes the chain.

Alkylating agents are used to treat leukemias, lymphomas, and Hodgkin's disease. Their main mode of action is to alkylate the guanine in DNA, which leads to abnormal base-pairing of the altered guanine with thymine and guanine rather than with cytosine. This DNA damage blocks cell division in the tumor as well as in normal cells.

Bleomycin, another antineoplastic drug, acts by fragmenting DNA.

Several other rare disorders caused by faulty DNA repair include ataxia telangiectasia, Bloom's syndrome, and Fanconi's anemia.

DNA Recombination

In contrast to DNA replication, DNA recombination involves the mixing of parental DNA from two different sources. DNA recombination is involved in:

1. Meiosis in eukaryotes.
2. Repair of DNA when both strands are damaged.

3. Integration of phage and plasmid DNA into bacterial host cells.
4. Transduction of DNA segments by viruses from a host cell into a newly infected cell.

Recombination is either general or site-specific. General recombination is confined to homologous (highly similar) DNA segments. The repair of a DNA segment when both strands are damaged involves general recombination—the DNA used in the repair process will come from an identical portion of another DNA molecule. Site-specific recombination is seen in the transduction of DNA segments by viruses from a host cell to a new host cell. The transfer of DNA between chromosomes is a form of site-specific recombination known as **transposition.** Mobile genetic materials such as the circular DNA of plasmids are called **transposons.**

Transcription in Prokaryotes

The synthesis of RNA from DNA is termed transcription. Transcription is followed by translation (protein synthesis from mRNA). Transcription creates three basic types of RNA: messenger RNA (mRNA), ribosomal RNA (rRNA), and transfer RNA (tRNA).

Transcription begins when RNA polymerase binds to promoter sites within the DNA template. Promoter sites in prokaryotes often have the base sequence TATAA at -10 (10 bases prior to the start of transcription) and TTGACA at -35.

Unlike DNA polymerase, RNA polymerase does not require a primer. RNA polymerase unwinds part of the double helical DNA, creating a template for transcription. RNA polymerase then begins synthesis of the RNA chain by adding NTPs in a 5' to 3' direction, creating a strand complementary to the DNA template. RNA chains usually begin with guanosine or adenosine triphosphate. RNA polymerase transcribes only one of the DNA strands at a time. As successive nucleotides are added to the RNA chain, the 5' end of this RNA retains its original triphosphate group, which provided the 5' terminus. Transcription ends when RNA polymerase encounters a chain termination signal. One such signal is a DNA segment rich in guanine and cytosine followed by a segment rich in adenine and thymine. This sequence leads to a hairpin bend in mRNA. The 3' end of mRNA consists of four or more uridine residues. Hydrogen bonding between mRNA and DNA is relatively weak, and around the hairpin bend, it is even weaker. As a result, the mRNA dissociates from the DNA template and the DNA rewinds.

Another mechanism for terminating transcription is chain terminating factors. In *E. coli* the rho protein, an ATPase, causes mRNA to dissociate from DNA.

Unlike DNA polymerase, RNA polymerase is not a nuclease and cannot therefore correct errors in the new polynucleotide chain. Hence, transcription has a higher error rate than DNA replication.

Rifampicin, an antibiotic used to treat tuberculosis, binds to bacterial RNA polymerase, which inhibits the initiation of bacterial RNA synthesis. Poisonous mushrooms contain amanitin, an inhibitor of human RNA polymerase. Amanitin blocks transcription at the elongation phase. Actinomycin D, used in cancer chemotherapy, intercalates between two GC base pairs in DNA, thereby inhibiting transcription.

Transcription in Eukaryotes

In prokaryotes, transcription and translation occur within the same cellular compartment. In fact translation of bacterial mRNA begins during transcription. In eukaryotic cells, transcription and translation occur in different cellular compartments: transcription occurs within the nucleus and translation outside the nucleus. Hence translation can occur only after transcription has finished. Human mitochondrial DNA is an exception to this general rule; mitochondrial DNA is transcribed within the mitochondria.

Another difference between prokaryotic and eukaryotic transcription is that prokaryotes translate primary transcripts (the RNA synthesized by RNA polymerase) whereas eukaryotes extensively process primary transcripts prior to translation (see next section).

In eukaryotic cells, promoter sites generally have the base sequence TATAA centered at base -25 (25 bases behind the base where transcription will begin). Additional promoter sequences are required such as GC and CAAT located farther from the start of transcription.

Eukaryotic cells have three types of RNA polymerase, in contrast to the single type found in prokaryotes. These RNA polymerases share many common features with prokaryotic RNA polymerase: they do not require a primer, they synthesize RNA in a 5' to 3' direction, and they lack nuclease activity to correct errors in transcription. RNA polymerase type I produces rRNA precursors, type II produces mRNA precursors, and type III produces tRNA precursors.

Transcription factors are proteins required to initiate transcription. A number of transcription factors operate in eukaryotic cells. Transcription factors recognize the promoter sites in DNA.

Eukaryotes also have DNA base sequences known as enhancers, which stimulate the transcription of genes at other sites often distant from the enhancer itself. Enhancers are expressed in cells requiring larger amounts of certain proteins. Thus B-lymphocytes have enhancers for immunoglobulin synthesis.

Transcription in both prokaryotes and eukaryotes generally starts with guanosine or adenosine triphosphate. During the transcription of eukaryotic mRNA, the 5' end is capped. Two changes occur in capping. First, the terminal guanine is methylated, forming 7-methylguanylate. Second, the ribose of residues 2 and 3 are methylated. Caps have two functions: they protect the 5' end of mRNA from nucleases and phosphatases, and they foster translation. Ribosomal and transfer RNA do not have caps.

The signals terminating transcription in eukaryotes are uncertain. The DNA sequence AAUAAA plays a part in termination. Endonuclease cleaves the RNA, releasing it from the DNA template.

Reverse Transcription

Oncogenic (tumor-producing) RNA viruses or retroviruses synthesize DNA from RNA, a reversal of the usual procedure, and insert this DNA, known as an oncogene, into the chromosomes of animal cells. This reverse transcription is catalyzed by an **RNA-directed DNA polymerase**.

Modification of RNA after Transcription

Following synthesis by RNA polymerase, the mRNA of prokaryotes requires little or no further modification to begin translation. Ribosomal RNA and transfer RNA of prokaryotes and eukaryotes, however, arise from larger RNA chains, which are cleaved and modified after transcription. A variety of nucleases cleave RNA into smaller segments, which will become tRNA and rRNA.

Nucleotides are added to various RNA molecules after transcription. Transfer RNA molecules lacking CCA at their 3' ends will receive these three nucleotides.

The bases and sugars of rRNA are sometimes modified after transcription. The 2' hydroxyl group of ribose can be methylated in human rRNA. Bases can be modified to create unusual bases such as pseudouridine.

Human tRNA is formed by a series of steps after transcription. Bases are removed from the leader on the 5' end, the intron is deleted, certain bases are modified, and the 3' UU is converted to CCA.

After eukaryotic mRNA is transcribed, poly A polymerase adds a tail of about 250 A residues to the 3' end. The role of this poly A tail is uncertain. The mRNA for histones lacks this poly A tail.

Perhaps the most important change to eukaryotic mRNA after transcription is splicing, which involves removal of its introns. The 5' end of introns contains GU while the 3' end has AG. Somewhere between the two ends of the intron is the branch site containing adenine. Splicing begins at the 5' end of the intron. The 2' hydroxyl group of adenine within the branch site of the intron forms a phosphodiester bond with the 5' phosphate of the intron. This releases the 5' end of the adjacent exon (exon 1), allowing the 3' end of that exon to form a phosphodiester bond with the 5' end of the next exon (exon 2). The bond between exon 2 and the intron is cleaved, releasing the intron and leaving exons 1 and 2 spliced. Spliceosomes built from ribonucleoproteins hold the exons together prior to this splicing.

There are rare examples of RNA molecules that can splice themselves.

The Genetic Code

Each DNA strand codes for the synthesis of many polypeptides. A segment of DNA that codes for one polypeptide chain is termed a **gene.** Each mRNA can code for one or several polypeptides.

Messenger RNA is functionally, not structurally, partitioned into sequential trinucleotide fragments called **codons.** Each codon leads to the addition of one particular amino acid during **translation,** or protein synthesis. There are 4^3, or 64, possible trinucleotide sequences of the four nucleotides in mRNA. Three codons—UAA, UAG, and UGA—do not represent amino acids, but rather they signal the point of chain termination. Each base is part of only one codon; hence the genetic code has no overlap. Likewise no bases separate codons.

Most amino acids have several codons, as shown in Figure 15-2. Hence, the genetic code is termed **degenerate;** i.e., redundancy exists among the code words for most amino acids. This degeneracy usually involves the third base of the codon, although occasionally the other two bases also vary for a given amino acid. Isoleucine, for example, has three codons that differ only in the third base: AUA, AUC, and AUU. Serine, on the other hand, has six codons that exhibit degeneracy in all three bases: AGC, AGU, UCA, UCC, UCG, and UCU.

Fig. 15-2 The codon dictionary. The chain-termination codons are indicated by "End." (Adapted from A. L. Lehninger, *Biochemistry*, 2nd ed. New York: Worth, 1975. Fig. 34–1.)

AAA Lys	CAA Gln	GAA Glu	UAA (End)
AAG Lys	CAG Gln	GAG Glu	UAG (End)
AAC Asn	CAC His	GAC Asp	UAC Tyr
AAU Asn	CAU His	GAU Asp	UAU Tyr
ACA Thr	CCA Pro	GCA Ala	UCA Ser
ACG Thr	CCG Pro	GCG Ala	UCG Ser
ACC Thr	CCC Pro	GCC Ala	UCC Ser
ACU Thr	CCU Pro	GCU Ala	UCU Ser
AGA Arg	CGA Arg	GGA Gly	UGA (End)
AGG Arg	CGG Arg	GGG Gly	UGG Trp
AGC Ser	CGC Arg	GGC Gly	UGC Cys
AGU Ser	CGU Arg	GGU Gly	UGU Cys
AUA Ile	CUA Leu	GUA Val	UUA Leu
AUG Met	CUG Leu	GUG Val	UUA Leu
AUC Ile	CUC Leu	GUC Val	UUC Phe
AUU Ile	CUU Leu	GUU Val	UUU Phe

AUG, the only codon for methionine, is the chain-initiation codon for proteins of higher organisms. The codons preceding AUG signal whether this codon is read as a chain initiator or a further amino acid.

Human mitochondria use different tRNA molecules than the rest of the cell. This may explain why the genetic code for mitochondrial proteins differs slightly from that of cellular proteins. The DNA strand used as a template for a given mRNA has base triplets that are complementary to the codons (they are not considered codons themselves). Any change in the DNA base sequence will be transmitted in both DNA replication and transcription. Changes in a single base, or point mutations, may involve base transition, transversion, deletion, or insertion.

In **transitional mutations,** a purine replaces another purine in DNA and a pyrimidine replaces another pyrimidine. 5-Bromouracil, which is not used medically, is a thymine analog that can replace thymine in DNA. Since it can base-pair with both adenine and guanine, a guanine, rather than the adenine that thymine would specify, can be incorporated into the complementary strand during DNA replication or transcription.

2-Aminopyrine, a drug whose medical use was abandoned years ago because of its toxic effect on bone marrow, can replace adenine (6-aminopurine) or guanine in DNA, and it, too, induces transitional mutations.

In **transversional mutations,** a purine replaces a pyrimidine in DNA and a pyrimidine replaces a purine. These replacements occur in pairs and sometimes may simply represent a transposition of the two bases; e.g., AT may be replaced by TA. Spontaneous mutations are often the result of transversions.

Both transitional and transversional mutations are **point** mutations in that they change only one codon in mRNA. A codon change from AAA to AAG, for instance, will do no harm, since both codons are read as the code for lysine. A change from AAA to ACA, however, will place threonine rather than lysine into the protein. This change in primary structure may or may not alter the secondary, tertiary, and quaternary structures of the protein or change its enzymatic behavior. In sickle-cell anemia, however, valine replaces glutamate at position 6 in the β-chain of hemoglobin, which results in severe distortions in secondary, tertiary, and quaternary structure of hemoglobin and the consequent sickling of erythrocytes.

Deletion mutations occur either as a result of the loss of a base or after chemical damage to a base, such as the alkylation of guanine with alkylating drugs. In the latter case, the damaged base cannot base-pair normally and might not be read during transcription.

Acridine causes **insertion mutations,** because it can fit between two successive bases and be read during transcription so that an additional base will be inserted into the complementary mRNA strand.

The deletion and insertion mutations, which also are point mutations, are far more detrimental than transition or transversion mutations, because they shift the triplet reading-frame. The deletion of A from the DNA below, for example, changes the triplets (not codons) from TCA, GTG, T . . . to TCG, TGT, . . . :

$$TCAGTGT \cdots \xrightarrow{\text{deletion of A}} TCGTGT \cdots$$

All DNA triplets occurring in the chain in the 3′ direction after such a **frame-shift mutation** will be thrown out of register, producing distorted mRNA codons. Frame-shift mutations near the 5′ end of a gene usually terminate polypeptide synthesis or produce a defective polypeptide, because most of the mRNA codons have been altered. Such mutations near the 3′ end of a gene, however, change relatively few mRNA codons and may produce functional polypeptides.

Any type of mutation may produce chain-terminating codons that prematurely stop polypeptide synthesis. Double or triple base changes within a triplet are rare compared to the point mutations previously discussed.

Recombinant DNA

Recombinant DNA technology is the process in which desired segments of DNA are selectively removed from a large DNA segement, purified, amplified, and produced in large quantities. The steps used in recombinant DNA technology are:

1. Excision of the desired DNA segment using restriction endonucleases, which cleave both DNA strands.
2. Separation of DNA fragments using gel electrophoresis.
3. Identification of DNA fragments.
4. Amplification of the desired DNA fragment.

A variety of methods have been employed to identify DNA fragments. In the Southern blotting technique DNA fragments are hybridized with a single-stranded ^{32}P-labelled DNA probe that matches a portion of the gene under investigation. Probes bind only to complementary DNA sequences. The position of the probe on autoradiography demonstrates which fragment has a complementary DNA sequence. Because the genetic code is degenerate, the amino acid sequence of a polypeptide does not itself specify the base sequence of the corresponding DNA. This poses a major problem when trying to construct DNA probes from knowledge of amino acid sequence. One way to circumvent this problem is to create probes for polypeptide sequences containing methionine and tryptophan, which are encoded by a single codon. Another way is to prepare DNA probes from the mRNA coding for a given protein using reverse transcriptase. Reverse transcriptase produces DNA complementary to the RNA template. DNA produced by reverse transcriptase is termed complementary or cDNA.

DNA fragments can also be identified by chemical cleavage in a process akin to the chemical sequencing of a polypeptide.

There are two general approaches to amplifying DNA: the **polymerase chain reaction (PCR)** and **cloning.** The PCR amplifies DNA far more rapidly than does the cloning method. The PCR proceeds as follows:

1. The DNA fragment to be amplified (call it X) is inserted into a plasmid vector whose DNA sequence is known.
2. The plasmid vector is heated to 90°C to separate the two DNA strands and then cooled, adding oligonucleotides complementary to the known vector DNA strands.
3. The oligonucleotides bind to the plasmid vector DNA on either side of the X fragment. They act as primers for DNA polymerase, which copies both DNA strands of X.
4. DNA X but not the plasmid vector DNA is copied. The resultant mixture is then heated and the cycle repeated many times until the requisite quantity of DNA X has been produced.

The second method for amplifying DNA is cloning. To clone a selected DNA fragment it must first be joined to a DNA vector, creating a recombinant DNA molecule. DNA vectors are molecules capable of independent replication within a host. Plasmids (circular duplex DNA molecules) and certain phages are ideal vectors for cloning DNA in bacteria. DNA ligase is used to join the fragment for cloning to the DNA vector. The recombinant DNA is then introduced into the host cell. The process of introducing foreign DNA into bacteria is termed **transformation.** Since most host cells will not take up external DNA, one could use specific host cells capable of incorporating foreign DNA. Alternatively, one could load the recombinant DNA into a viral genome. The virus then infects the host cells, introducing the recombinant DNA.

Bacteria can produce certain human proteins such as insulin. They cannot, however, produce proteins requiring post-translational modification. Hence eukaryotic host cells must be employed to synthesize such proteins.

To treat genetic diseases by gene replacement, recombinant DNA containing the desired gene can be introduced in vitro into host cells. This process is known as **transfection.** Among the methods for transfection are:

1. Exposure to calcium phosphate, which promotes DNA uptake by host cells.
2. Retroviruses. These RNA tumor viruses use reverse transcriptase to synthesize DNA from their RNA. After infecting host cells, this viral DNA is inserted into the host genome.
3. Injection of foreign DNA into host cells.

Transfection occurs in the minority of host cells—most do not express the recombinant DNA. To help select cells that have been transfected, marker genes may be transfected along with the gene of interest, a process termed **cotransfection.** Only cells expressing the marker gene are selected.

Gene Structure

The genes of prokaryotes are continuous DNA sequences. In contrast, few human genes are continuous DNA sequences. Continuous human genes include those for most tRNA molecules and histones. Most human genes are discontinuous, with two or more expressed sequences called **exons** separated by untranslated regions called **introns**. The total length of introns far exceeds that of exons. Introns are often spliced out of RNA prior to translation. Exons often code specific portions of proteins. Hence new proteins can be built from exchange or duplication of previous exons.

Because bacteria cannot splice out introns, they cannot create intact genes from discontinuous human genes. To circumvent this problem, one can use reverse transcriptase to copy the mRNA for the required protein. The copy will have a complementary base sequence to the mRNA. This DNA copy is then taken up by a vector and transferred to a bacterial host cell.

Problems

Problem 1

Choose the incorrect statement about DNA polymerase.

  A. It edits as it synthesizes.
  B. DNA polymerase III is a highly processive enzyme.
  C. It cannot replicate DNA unless an RNA primer is first created.
  D. The energy to create phosphodiester bonds comes from hydrolysis of nucleoside triphosphates to monophosphates and subsequent hydrolysis of pyrophosphate.
  E. It synthesizes the lagging DNA strand continuously.

Problems 2–5

Match the enzymes below to the processes in Problems 2–5.

  A. DNA ligase
  B. Excinuclease
  C. DNA polymerase
  D. RNA polymerase

2. Cleaves DNA to excise thymine dimers.
3. Removes the RNA primer on DNA.
4. Joins Okazaki fragments.
5. Synthesizes RNA primer during DNA replication.

Problem 6

What is the result of a transversion mutation in the coding strand of DNA from TAA to TAC? You may refer to Figure 15-2. Select one answer.

  A. Codon changes from AUU to AUG; methionine replaces isoleucine.
  B. Codon changes from UUA to GUA; valine replaces leucine.

C. Frame-shift occurs.

D. DNA replication stops.

Problem 7

Which of the following amino acid substitutions could result from a single base mutation in DNA (rather than a double or triple base change in the same triplet)? You may refer to Figure 15-2. Select one or more answers.

A. Leucine to lysine

B. Phenylalanine to lysine

C. Phenylalanine to leucine

D. Isoleucine to leucine

Problem 8

Place the following steps in DNA replication for the lagging strand in order.

A. DNA polymerase adds NTPs in a 5′ to 3′ direction.

B. Primase creates an RNA primer.

C. DNA ligase joins the DNA fragments.

D. Unwinding proteins open the DNA double helix.

E. DNA polymerase fills the gap between Okazaki fragments and removes the RNA primer.

Answers

1. E.

2. B.

3. C.

4. A.

5. D. Primase is the RNA polymerase that creates the RNA primer complementary to each DNA strand.

6. B. During transcription the DNA triplet TAA binds to mRNA as follows:

DNA	5′ T—A—A 3′
mRNA	3′ A—U—U 5′

By convention, polynucleotide sequences are written 5′ to 3′. Hence the mRNA codon is UUA, not AUU. Similarly TAC corresponds to GUA.

7. C, D. A single base change converts the code for Phe (UUC or UUU) to one for Leu (UUA, UUG, CUU, or CUC). Similarly, the code for Ile (AUU or AUC) can be converted to CUU or CUC for Leu. The other two amino acid mutations require double and triple base changes, respectively.

8. D, B, A, E, C.

References

Devlin, T. M. *Textbook of Biochemistry with Clinical Correlations* (3rd ed.). New York: Wiley–Liss, 1992. Pp. 648–677, 695–719, 768–801.

Mathews, C. K., and van Holde, K. E. *Biochemistry*. Redwood City, Calif.: Benjamin/Cummings, 1990. Pp. 817–883, 910–948.

Murray, R. K., Granner, D. K., Mayes, P. A., and Rodwell, V. W. *Harper's Biochemistry* (22nd ed.). Norwalk, Conn.: Appleton & Lange, 1990. Pp. 366–394, 424–439.

Stryer, L. *Biochemistry* (3rd ed.). New York: Freeman, 1988. Pp. 78–140, 649–685, 687–701, 703–732.

Translation

Protein synthesis from mRNA is termed translation because the four-letter base code is converted to the codes for many more amino acids. Translation has three stages: initiation, elongation, and termination.

Initiation

Peptide bond formation is endergonic and must therefore be coupled with an energy-producing reaction. The activation of amino acids prior to peptide bond formation accomplishes this aim. Each amino acid has its own aminoacyl-tRNA synthetase that activates it and then joins it to its particular tRNA. This two-step reaction consumes two high-energy phosphate bonds because it converts ATP to AMP and PP_i. The amino acid binds to the 3' end of tRNA, which has the trinucleotide sequence CCA.

$$\text{Amino acid} + \text{ATP} \rightleftharpoons \text{aminoacyl-AMP} + PP_i$$

$$\text{Aminoacyl-AMP} + \text{tRNA} \rightleftharpoons \text{aminoacyl-tRNA} + \text{AMP}$$

Net: Amino acid + ATP + tRNA $\rightleftharpoons$ aminoacyl-tRNA + AMP + PP_i

The activation of amino acids resembles that of fatty acids.

Aminoacyl-tRNA synthetases must accurately recognize both the amino acid they activate and the tRNA specific for that amino acid. Transfer RNA is more easily recognized than amino acids because each tRNA has its own base sequence. Related amino acids, however, may have only slight structural differences, leading to errors in recognition. To counteract this problem, aminoacyl-tRNA synthetase corrects errors in activation by hydrolyzing incorrect aminoacyl-AMP molecules prior to transfer to tRNA. This form of editing is similar to that of DNA polymerase.

Transfer RNA

Transfer RNA is a single RNA chain with 73 to 93 bases folded into a cloverleaf pattern. About 10% of the bases are unusual ones such as methylcytidine and inosine. The four arms of the cloverleaf consist of folded chains. Hydrogen bonding between GC and AU holds these folds together. Indeed about half of the bases in tRNA are paired.

Transfer RNA is synthesized as larger precursors. The primary transcripts are cleaved by a variety of ribonucleases. The RNA rather than the protein in ribonuclease P acts as an enzyme. Post-transcriptional base modification follows this cleavage.

The 3′ and 5′ ends of tRNA are found in the same arm of the cloverleaf. The 3′ end contains the base sequence CCA, which serves as the amino acid attachment site. The 5′ end is phosphorylated. Guanine is the most common 5′ terminal base.

The opposite arm of the cloverleaf contains the anticodon. Hence the amino acid on aminoacyl-tRNA is distant from the anticodon. The anticodon loop contains seven bases in set order shown below.

5′ Pyrimidine–pyrimidine—trinucleotide anticodon–modified purine–other base 3′

The anticodon is a trinucleotide sequence complementary to the codon for the specified amino acid. Some anticodons can bind to more than one codon. The general rules that govern binding of anticodons to codons are as follows:

1. The first base in the anticodon (corresponding to the third base in the codon) determines whether the anticodon can bind to one, two, or three codons. Thus, if A or C is the first base, the anticodon can bind to only one codon; if G or U, it can bind to two; and if inosine, it can bind to three.
2. The first two bases in a codon are specifically paired. A single tRNA can recognize only one combination of these bases.

The amino acid carried by tRNA plays no role in the selective binding of tRNA to codons.

Ribosomes

Ribosomes are ribonucleoproteins built from smaller subunits. The ribosomes of prokaryotes have a sedimentation coefficient of 70S. The larger subunit is 50S while the smaller subunit is 30S. The subunits, in turn, are built from rRNA plus a variety of proteins. The larger subunit contains 23S and 5S rRNA, whereas the smaller subunit contains 16S rRNA.

Eukaryotic ribosomes differ in size from those in prokaryotes. Eukaryotic ribosomes in the cytosol are 80S, with subunits of 60S and

40S. The 60S subunit contains 5S, 5.8S, and 28S rRNA. The 40S subunit contains 18S rRNA. Mitochondrial ribosomes are 55–60S.

Ribosomal RNA is a highly folded molecule. Like other RNA, the folds are linked by hydrogen bonding between base pairs.

Provided all the RNA and protein components are present, ribosomes can assemble themselves.

Initiation of Polypeptide Chain Formation

Translation occurs in a 5′ to 3′ direction. Polypeptide chains are built starting from their amino terminal ends.

Formylmethionine initiates bacterial protein synthesis. To create formylmethionine (fMet), a tRNA specific for formylmethionine first binds to methionine via aminoacyl-tRNA synthetase. Formyl-tetrahydrofolate then donates its formyl group to methionine. The anticodon on fMet-tRNA recognizes the initiating codon on mRNA. The codon most often initiating translation is AUG; GUG can also start the process.

To initiate protein synthesis in bacteria, three protein initiation factors (IF1, IF2, and IF3) must first bind to the 30S ribosomal subunit while GTP binds to IF2. Messenger RNA and fMet-tRNA then bind to this complex and IF3 is released. Next the 50S ribosomal subunit binds to this complex with release of IF1 and IF2; this is coupled with the hydrolysis of the bound GTP to GDP and P_i. The net result is a 70S initiation complex with fMet at the peptidyl site (or P site) of the ribosome.

Polypeptide Chain Elongation

Elongation begins with the binding of an aminoacyl-tRNA to the vacant aminoacyl site (or A site) on the ribosome. Only the aminoacyl-tRNA with anticodon matching the codon at the A site can bind. The EF proteins are elongation factors. EF-Tu is the elongation factor that brings the appropriate aminoacyl-tRNA to the A site. The hydrolysis of GTP then releases EF-Tu from the aminoacyl-tRNA, which remains bound to the ribosome. A second elongation factor, EF-Ts, binds to EF-Tu. The GDP attached to EF-Tu is replaced by GTP and EF-Ts is released. EF-Tu bound to GTP then picks up another aminoacyl-tRNA for delivery to the A site.

There is a delay between the arrival of an aminoacyl-tRNA at the A site and peptide bond formation with that amino acid. During this period, aminoacyl-tRNAs that do not have the correct anticodon are prone to detach from mRNA. In this way translation is edited.

Peptidyl transferase is an enzyme of the 50S ribosomal subunit that forges peptide bonds between amino acids bound to tRNA. The energy driving this reaction comes from the previous activation of amino acids into aminoacyl-tRNA.

$$
\begin{array}{c}
\text{O} \qquad\quad \text{R}_1 \quad\ \text{O} \\
\|\qquad\qquad\ |\qquad\ \| \\
\text{HC—NH—CH— C—O—tRNA}
\end{array}
\qquad\qquad
\begin{array}{c}
\text{O} \qquad\quad \text{R}_2 \quad\ \text{O} \\
\|\qquad\qquad\ |\qquad\ \| \\
\text{HC—NH—CH—C —O—tRNA}
\end{array}
$$

fMet-tRNA at P site   Aminoacyl-tRNA at A site

Peptidyl transferase

$$
\begin{array}{c}
\text{O} \qquad\ \text{R}_1 \quad \text{O} \qquad\quad \text{R}_2 \quad\ \text{O} \\
\|\qquad\ |\qquad\ \|\qquad\qquad |\qquad\ \| \\
\text{HC—NH—CH—C —NH—CH—C —O—tRNA}
\end{array}
$$

HO—tRNA

tRNA at P site   Peptidyl-tRNA at A site

Thus after peptide bond formation an uncharged tRNA occupies the P site and a peptidyl-tRNA occupies the A site. To continue elongation these components must be shifted in a process termed **translocation.** During translocation the uncharged tRNA leaves the P site, the peptidyl-tRNA moves from the A to the now-vacant P site, and the ribosome advances three bases to read the next codon. Elongation factor G (EF-G or translocase) is required for translocation. Like the other elongation factors, EF-G binds to GTP. Following translocation, the hydrolysis of GTP leads to release of EF-G from the ribosome. Thus, translocation requires the expenditure of one GTP per peptide bond.

In prokaryotes a single ribosome translates each mRNA. Among eukaryotes, a number of ribosomes ordinarily translate a single mRNA molecule at the same time. This complex of ribosomes bound to a single mRNA is called a **polysome.** Each ribosome acts on its own, synthesizing a complete polypeptide.

Termination of Translation

The codons UAG, UAA, and UGA are stop signals. No aminoacyl-tRNA molecules will bind to these codons. Instead, the so-called releasing factors recognize stop signals. When a releasing factor binds to a stop signal, peptidyl transferase hydrolyzes the bond joining the polypeptide at the A site to tRNA. Suddenly the polypeptide, tRNA, and mRNA are released and the ribosome splits into its subunits.

Four high-energy phosphate bonds are consumed to synthesize each peptide linkage: two from ATP are used during amino acid activation, one from GTP during aminoacyl-tRNA binding to the A site, and another from GTP during translocation.

Antibiotics and Protein Synthesis

Erythromycin binds to the 50S subunit of bacterial ribosomes, inhibiting translocation. Chloramphenicol also binds to the 50S subunit; it inhibits peptidyl transferase.

Aminoglycosides and tetracyclines bind to the 30S subunit of bacterial ribosomes, blocking aminoacyl-tRNA binding.

Streptomycin prevents the binding of fMet-tRNA to ribosomes, thereby blocking chain initiation. It also causes misreading of the mRNA.

Protein Synthesis in Eukaryotes

Methionine rather than formylmethionine initiates protein synthesis in eukaryotes. Like prokaryotes, a special tRNA is used for initiation; methionine added later in the chain is carried by another tRNA.

A single codon, AUG, closest to the 5′ end of mRNA, initiates protein synthesis in eukaryotes. Eukaryotic mRNA has only one start signal per molecule and can therefore code for only one polypeptide; prokaryotic mRNA can have multiple start signals, allowing it to code for more than one polypeptide.

Eukaryotes have more initiating factors and elongating factors than prokaryotes. In general these factors use GDP in the same way as prokaryotic initiation and elongation factors. Certain eukaryotic cells have kinases that can phosphorylate initiating factor 2, the protein that carries Met-tRNA to the P site. After phosphorylation, this initiating factor becomes inactive. Kinases are used to regulate protein synthesis. For example, in the absence of heme, kinases inhibit globin synthesis.

Diphtheria toxin is a powerful inhibitor of human protein synthesis. The toxin contains an enzyme that transfers ADP-ribose from NAD to elongation factor 2, thereby blocking translocation.

Intracellular Protein Sorting and Post-Translational Modification

All proteins except those coded in mitochondrial and chloroplast DNA are synthesized on ribosomes in the cytosol. Many proteins require further modification and must be transferred to other cellular structures, particularly the endoplasmic reticulum (ER).

Only polypeptides bearing certain sequences on their amino terminal ends will cross the ER membrane. Signal recognition particle (SRP) is a ribonucleoprotein that detects these signals soon after synthesis. SRP binds to cytoplasmic ribosomes containing such growing polypeptides, and chain elongation is slowed as the SRP-ribosome complex moves to the ER membrane. SRP then binds to an SRP receptor on the ER membrane. Following binding, SRP is released. Chain elongation then resumes at full rate. The growing polypeptide chain enters the ER lumen as a relatively straight chain.

Inside the ER lumen many proteins are modified. Disulfide bonds are formed by oxidizing the SH groups of cysteine. Amino terminal sequences, including the N-terminal methionine of eukaryotes, are often cleaved by signal peptidases.

Glycoprotein synthesis by glycosylation of polypeptides begins within the ER. Glycoproteins may be O-linked or N-linked. In O-linked glycoproteins the OH in serine or threonine is joined to the carbohydrate. In N-linked glycoproteins the carbohydrate is joined to the amide group of glutamine.

Dolichol phosphate, an activated lipid carrier, is required to transfer oligosaccharides during synthesis of N-linked glycoproteins.

The Golgi complex consists of a series of membranous envelopes or cisternae. The Golgi complex is a major protein-sorting apparatus that also modifies the carbohydrate elements of glycoproteins. This process of terminal glycosylation follows the core glycosylation occurring within the ER. This complex has three functional areas: cis, medial, and trans. The cis Golgi, nearest to the ER, receives proteins in transfer vesicles. Within the cis Golgi, mannose is removed from certain glycoproteins. Proteins packaged in vesicles move from the cis Golgi to the medial Golgi and from there to the trans Golgi. Within the medial Golgi, fucose and *N*-acetylglucosamine are added and further mannose is removed from glycoproteins. Sialic acid and galactose are added to glycoproteins in the trans Golgi.

If mannose-6-phosphate is added to glycoproteins within the Golgi complex, the glycoproteins will be transported from the Golgi complex to the lysosomes. Failure to produce mannose-6-phosphate due to a deficiency in phosphotransferase in I-cell disease results in a selective deficiency of eight lysosomal enzymes, which are present in other parts of the cell. This lysosomal enzyme deficiency leads to retardation and skeletal deformities.

The majority of mitochondrial proteins are encoded by nuclear rather than mitochondrial DNA. Hence most are synthesized in the cytosol and then transported to the mitochondria. Proteins destined for mitochondria have a mitochondrial entry sequence that acts as a signal. There are different sequences according to the destination within the mitochondria. Proteins heading for the outer membrane are most easily delivered to their correct destination. Proteins heading for the inner membrane, the intermembrane space, and the matrix have a more complex path to follow. Because of their charge, a transmembrane potential is required for passage through the outer and inner mitochondrial membranes.

Proteins such as DNA polymerase, RNA polymerase, and histones are synthesized in the cytosol and then transported into the nucleus via nuclear pores.

Hormonal Control of Protein Synthesis

Hormones that stimulate protein synthesis include insulin, the androgens, thyroid hormones, somatomedins, and growth hormone (STH). Insulin stimulates tissue amino acid uptake and protein synthesis, while inhibiting gluconeogenesis. Androgens such as testosterone promote a positive nitrogen balance and are responsible for the larger muscle mass of males compared to females. They stimulate RNA polymerase and increase the rate of transcription. Thyroid hormones and growth hormone markedly stimulate transcription and protein synthesis. When either is deficient, normal growth cannot occur.

The glucocorticoids lead to protein catabolism in muscle. They also stimulate gluconeogenesis in the liver.

Problems

Problem 1

Put the following steps in prokaryotic translation into the proper sequence:

A. 50S ribosomal subunit binds to 30S subunit, mRNA, and tRNA.
B. Ribosome advances three bases to read next codon.
C. Initiating factors bind to 30S ribosomal subunit.
D. Releasing factors bind to stop signal, causing peptidyl transferase to hydrolyze the bond joining the polypeptide at the A site to the tRNA.
E. mRNA and fMet-tRNA bind to the 30S ribosomal subunit.
F. Uncharged tRNA leaves the P site, and the peptidyl-tRNA from the A site moves to the now-vacant P site.
G. Peptidyl transferase forges a peptide bond.
H. Elongation factors bring aminoacyl-tRNA to bind to the A site.

Problem 2

Which step in translation does *not* consume a high-energy phosphate bond?

A. Translocation
B. Amino acid activation
C. Peptidyl transferase reaction
D. Binding of aminoacyl-tRNA to the A site, with subsequent release of elongation factor.

Problem 3

During translation in eukaryotic cells:

A. The initiating amino acid is formylmethionine.
B. Each mRNA codes for several polypeptides.
C. Unusual bases can be added.
D. Initiating factors require ATP.
E. Regulation may occur by kinases that phosphorylate initiating factors.

Problems 4–8

Match the organelles below to the features in Problems 4–8:

A. Golgi complex
B. Endoplasmic reticulum
C. Mitochondria
D. Lysosomes

4. Dolichol phosphate used to create N-linked glycoproteins.
5. Mannose-6-phosphate mediates transport of polypeptides to this organelle.
6. SRP brings polypeptides to this organelle.
7. Site of terminal glycosylation.
8. Transmembrane potential required for bringing polypeptides into this organelle.

Answers

1. C, E, A, H, G, F, B, D.
2. C.
3. Formylmethionine is the initiating amino acid in prokaryotic translation. In prokaryotes each mRNA codes for several polypeptides. In eukaryotes each mRNA codes for one polypeptide. Unusual bases are added by post-translational modification. The initiating factors use GTP, not ATP.
4. B.
5. D.
6. B.
7. A.
8. C.

References

Devlin, T. M. *Textbook of Biochemistry with Clinical Correlations* (3rd ed.). New York: Wiley–Liss, 1992. Pp. 723–763.

Mathews, C. K., and van Holde, K. E. *Biochemistry.* Redwood City, Calif.: Benjamin/Cummings, 1990. Pp. 954–988.

Murray, R. K., Granner, D. K., Mayes, P. A., and Rodwell, V. W. *Harper's Biochemistry* (22nd ed.). Norwalk, Conn.: Appleton & Lange, 1990. Pp. 395–407.

Stryer, L. *Biochemistry* (3rd ed.). New York: Freeman, 1988. Pp. 733–798.

Regulation of Gene Expression

This chapter examines genes and their regulation. Prokaryotic gene expression will be discussed first, followed by examination of chromosomes and eukaryotic gene expression.

Prokaryotic Genes

Certain bacterial genes are grouped into **operons.** Operons include related genes that are controlled by the same signals. Operons function by two processes: **induction** and **repression.** Induction is the turning on of transcription of a structural gene due to the presence of an inducer. Repression is the turning off of transcription of a structural gene due to the presence of a repressor. The lactose operon of *Escherichia coli* consists of two basic elements: **structural genes** for the proteins it produces and **control sites,** composed of an **operator** and a **promoter. Regulator genes** governing operon function lie adjacent to the operon.

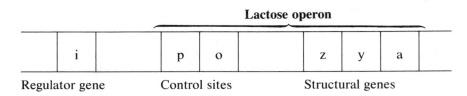

The lactose operon has three structural genes or cistrons, which are transcribed onto a single polygenic mRNA. Gene z codes for β-galactosidase, which hydrolyzes lactose to glucose and galactose. The y gene codes for galactoside permease, which transports lactose into *E. coli*. Gene a codes for thiogalactoside transacetylase, an enzyme that acetylates thiogalactosides. The role of acetylation in lactose metabolism is uncertain.

The regulator gene sits adjacent to but outside the operon. Regulator genes produce proteins called **repressors,** which bind to the operator site of their respective operon. The presence of repressors and the absence of inducers turns off transcription.

When *E. coli* is initially exposed to lactose, it has only a small amount of these enzymes. Nevertheless the available β-galactosidase converts some of the lactose to another disaccharide, allolactose. Allolactose, in turn, induces the synthesis of the three enzymes coded by the lactose operon.

Inducers such as allolactose bind to repressors, creating inducer-repressor complexes that will not attach to the operator site on the operon. By turning off repression, inducers selectively stimulate protein synthesis.

Catabolite repression is the process by which catabolism of one substrate represses that of other substrates. In *E. coli* the presence of glucose represses catabolism of other substrates such as lactose and galactose. Hence *E. coli* uses all available glucose prior to catabolizing other substrates. Cyclic AMP mediates catabolite repression.

Cyclic AMP stimulates transcription of many different genes in bacteria. Cyclic AMP levels in *E. coli* are low when glucose is abundant but rise during glucose deficiency. Cyclic AMP is therefore a signal of starvation. Cyclic AMP binds to **CAP (catabolite gene-activating protein)** and the complex binds to promoter sites in various catabolic operons (i.e., operons coding for catabolic enzymes). Once bound to mRNA, this complex creates an additional foothold for RNA polymerase to transcribe the structural genes. Thus, inducible catabolic operons such as the lactose operon are controlled by both their specific inducers and by the cyclic AMP–CAP complex.

The **tryptophan operon** of *E. coli* functions differently from the lactose operon. *E. coli* synthesizes tryptophan when it is not available in sufficient quantity in the environment. Tryptophan biosynthesis from chorismate requires three enzymes formed from five polypeptides. The structure of the tryptophan operon is shown below:

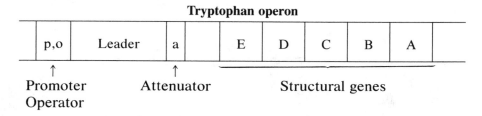

Control sites

The first step of tryptophan biosynthesis is the conversion of chorismate to anthranilate by anthranilate synthetase. Anthranilate synthetase is comprised of polypeptides transcribed from structural genes d and e of the tryptophan operon. Transcription determines the amount of this enzyme but feedback inhibition determines the activity of enzyme molecules after their synthesis.

Unlike the lactose operon, which must be induced to become active, the tryptophan operon is normally switched on; it must be repressed to be turned off. The tryptophan repressor gene, which is not part of the tryptophan operon, produces a protein termed the **tryptophan repressor.** The tryptophan repressor, when joined to tryptophan, binds to the operator site of the tryptophan operon, preventing

RNA polymerase from binding to the promoter site and thereby repressing synthesis of the structural genes. Thus, tryptophan serves as a corepressor in this process. The tryptophan repressor is not active in the absence of tryptophan, the essential corepressor.

Another difference between the tryptophan and the lactose operons is that the tryptophan operon has an attenuator site. The attenuator site is located within the 162-nucleotide leader between the promoter site and the first structural gene (e). The leader peptide contains two adjacent tryptophan residues. When tryptophan is present transcription continues past the codons for the two tryptophan residues, reaching a termination codon. This blocks further transcription of the structural genes. In the absence of tryptophan there is insufficient tryptophanyl-tRNA to transcribe the leader sequence past the codons for the two adjacent tryptophan residues. Hence transcription never reaches the chain-terminating codon and can continue to the structural genes: a polycistronic mRNA with the five structural genes is produced, thereby promoting tryptophan biosynthesis.

Structural genes can suppress other structural genes. *Salmonella*, for example, has two types of flagellar proteins, H1 and H2. To evade antibodies from host organisms, *Salmonella* switches periodically between these two proteins. Survival demands a complete change, leaving no trace of the former flagellar protein. The mRNA for H2 also codes for an H1 repressor. Translation of this repressor turns off H1 synthesis, allowing unopposed H2 production. To turn H1 synthesis back on, a recombinase inverts part of the H2 promoter sequence. The inverted sequence now runs 3' to 5' rather than 5' to 3'. When transcribed, this inverted promoter is ineffective. H2 synthesis stops due to lack of a promoter, and H1 synthesis resumes due to lack of a repressor. In time the recombinase enzyme inverts this promoter back to its original state and the cycle repeats itself.

Chromatin and Chromosomes

Free eukaryotic DNA is too long to fit into the nucleus; it must be condensed or compacted into chromatin. Almost all of the nuclear DNA in eukaryotes is bound to histones in nucleosomes that are organized as chromatin. Histones are basic proteins as a result of their high lysine and arginine content. There are five types of histones: H1, H2A, H2B, H3, and H4. Histones account for nearly half of the weight of eukaryotic chromosomes.

The genes encoding histones are unusual in that they do not have introns. Also unusual is the fact that mRNA for histones lacks poly A tails.

Nucleosomes consist of a DNA core and a linking region. The core is comprised of 140 base pairs of DNA encircling a histone octomer (two molecules each of histones H2A, H2B, H3, and H4). Thus histones occupy the center of the nucleosome core with DNA wrapped around the outside. The linking region of nucleosomes consists of 60 base pairs of DNA bound to one molecule of H1.

Chromatin is built from repeating nucleosome units. Chromatin is arranged like a string of beads: the string representing DNA and the beads representing globular proteins. For replication or transcription to occur these proteins must first dissociate from the DNA. Following this DNA must bind histones and reassemble into chromatin.

During the cell cycle chromatin becomes even more compacted into chromosomes, which are arranged in bands.

DNA replication in eukaryotes employs many replication forks for each DNA molecule. This collaboration greatly hastens DNA replication. During DNA replication the previous histones remain on the DNA duplex containing the leading strand, whereas new histones bind to the duplex with the lagging strand.

Nearly one-third of human DNA consists of base sequences repeated many times. Some of these sequences are genes while others are not. For example, the so-called Alu sequence of 300 base pairs appears almost one million times in the human genome; Alu may act as initiation sites for DNA replication rather than as a gene. Genes for rRNA are repeated more than a hundred times. Some genes that are repeated many-fold code for proteins or RNA required in high amounts. Thus, cells require large numbers of ribosomes and therefore benefit from having many copies of genes coding for ribosomal components. However, some proteins required in large amounts have only one copy of their respective gene. Reticulocytes, for example, produce large amounts of hemoglobin yet have few copies of the β-globin gene. Single genes may be duplicated many-fold in a process termed gene amplification.

The vast majority of the human genome does not code for proteins or RNA. Only a small fraction of human DNA is transcribed.

Mitochondrial DNA

Human mitochondria contain circular DNA, which codes for 13 proteins, 2 rRNAs, and 22 tRNAs. Nuclear DNA, in contrast, codes for thousands of proteins, many rRNAs, and 61 tRNAs. The genetic code for mitochondrial DNA differs slightly from that of nuclear and prokaryotic DNA. The UGA codon, for example, is not a chain terminator; instead it codes for tryptophan in mitochondria. AGG and AGA are chain terminators in mitochondria, whereas in other DNA they code for arginine.

Regulation of Gene Expression

The primary means of controlling gene expression in eukaryotes and prokaryotes is transcription. Genes active in transcription have fewer methylated cytosine bases than inactive genes. These methyl groups may block transcription by promoting formation of Z-DNA rather than B-DNA. Likewise the methyl groups may block binding of transcription factors.

Transcription factors are proteins that bind to regulatory sites within genes. Transcription factor IIIA is an unusual protein containing nine zinc ions. Each zinc ion is bound to two cysteine and two histidine residues and lies in the center of a finger-shaped polypeptide chain. The tip of each finger binds to five base pairs on the DNA. Transcription factor IIIA remains bound to certain genes during transcription.

The homeo box is a base sequence found in many species. This sequence is associated with genes governing development.

Problems

Problem 1

Choose the incorrect statement about the lactose operon of *E. coli*.

    A. Induction leads to the synthesis of β-galactosidase, galactoside permease, and thiogalactoside transacetylase.
    B. When repressed, lactose cannot readily enter *E. coli*.
    C. Repressors are produced from regulatory genes.
    D. Allolactose, an inducer, binds to repressors.
    E. Lactose is an important inducer of the lactose operon.

Problem 2

The galactose operon of *E. coli* has three structural genes for the enzymes that convert galactose to glucose-1-phosphate. Based on your understanding of the lactose and tryptophan operons you could predict that:

    A. This operon will normally be switched on but can be turned off by repression similar to the tryptophan operon.
    B. Low levels of cyclic AMP would activate transcription of the structural genes.
    C. The presence of glucose leads to induction of the galactose operon.
    D. Each structural gene would be transcribed on its own mRNA.
    E. An attenuator would block transcription in the presence of glucose-1-phosphate.

Problem 3

The nuclear DNA of eukaryotes differs from prokaryotic DNA in that:

    A. Some codons specify different amino acids.
    B. It lacks introns.
    C. It contains more repetitive DNA sequences.
    D. It does not form nucleosomes.
    E. It is generally circular.

Problem 4

The mechanism for cyclic alternation between synthesis of flagellar proteins H1 and H2 in *Salmonella* includes:

A. Cyclic AMP.
B. The mRNA for H2 codes for a repressor for H1.
C. The lack of critical amino acids during translation of a leader peptide for H2 promotes further H2 synthesis.
D. Recombinase inverts part of the promoter sequence for H1.

Answers

1. E. Lactose itself does not serve as an inducer.
2. B. The galactose operon is normally switched off, similar to the lactose operon. Catabolite repression blocks transcription when cyclic AMP levels are high. All three structural genes would be transcribed on a single mRNA. One would not expect an attenuator site for a catabolic pathway.
3. C. The genetic code for nuclear and prokaryotic DNA is the same; mitochondrial DNA has some differences. Nuclear DNA has large quantities of introns. Prokaryotic DNA is generally circular.
4. B.

References

Devlin, T. M. *Textbook of Biochemistry with Clinical Correlations* (3rd ed.). New York: Wiley-Liss, 1992. Pp. 805–845.

Mathews, C. K., and van Holde, K. E. *Biochemistry*. Redwood City, Calif.: Benjamin/Cummings, 1990. Pp. 996–1032.

Murray, R. K., Granner, D. K., Mayes, P. A., and Rodwell, V. W. *Harper's Biochemistry* (22nd ed.). Norwalk, Conn.: Appleton & Lange, 1990. Pp. 408–423.

Stryer, L. *Biochemistry* (3rd ed.). New York: Freeman, 1988. Pp. 799–850.

Metabolism and Its Hormonal Control

Metabolism is controlled primarily by the activities of enzymes at critical points in degradative and biosynthetic pathways. Enzyme activity, in turn, is regulated by the rates of enzyme synthesis and degradation, proteolytic activation (where appropriate), covalent modification, allosteric control, and regulatory proteins. The following diagrams illustrate four important metabolic states: immediately after a meal, early starvation (such as 12 hours after a meal), late starvation (as in a 4-week fast), and anaerobic (as in a person 2 minutes after cardiorespiratory arrest). The dark, thicker arrows show the predominant metabolic pathways. Decide which hormone(s) promote(s) each numbered reaction. The answers are at the end of the chapter.

After a meal, synthesis of protein, glycogen, and triglyceride predominates. The HMP shunt must be active to supply NADPH for fatty acid and cholesterol synthesis.

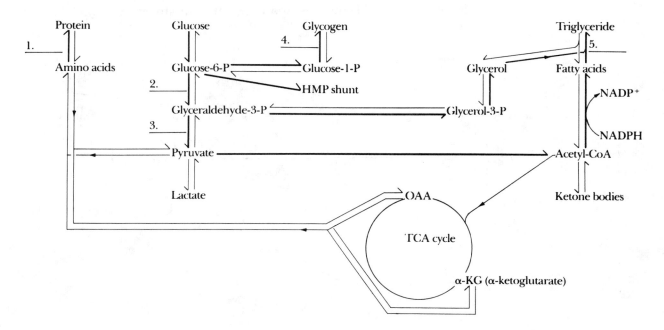

In **early starvation,** glycogenolysis and lipolysis are brisk. As the HMP shunt slows down, gluconeogenesis and ketone body formation start becoming active.

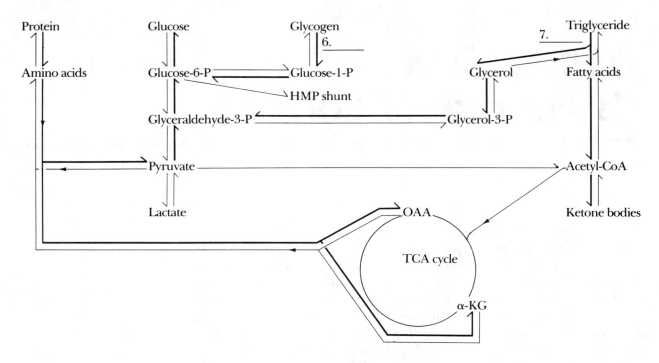

In **late starvation** (e.g., 4-week fast), catabolism of protein and tri-glyceride must make up for the loss of all glycogen reserves. Ketone body formation is brisk as the brain and heart utilize ketone bodies to supplement the increasingly scarce glucose. The main energy source for liver and muscles shifts from glucose to fatty acids. Utilization of ketone bodies and fatty acids spares protein, thereby slowing catabolism of muscles and other protein stores. The HMP shunt is shut down.

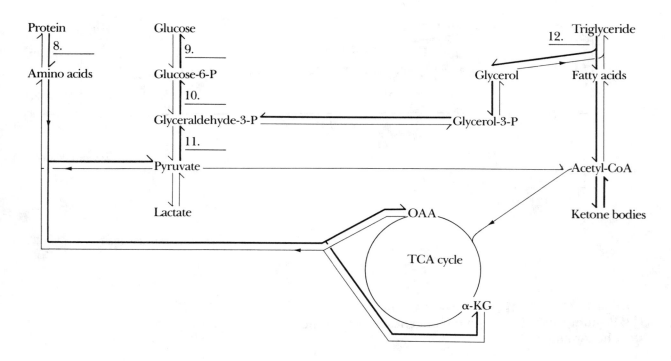

Anaerobic metabolism occurs when oxidative metabolism is limited or insufficient. During anaerobic metabolism, the NADH generated from glycolysis cannot be reconverted to NAD^+ via oxidative phosphorylation, which is shut down. Instead, the reduction of pyruvate to lactate recycles NAD^+, allowing glycolysis to continue. The TCA cycle is also shut down. The HMP shunt continues because the NADPH it generates can be recycled to $NADP^+$ during fatty acid and cholesterol synthesis.

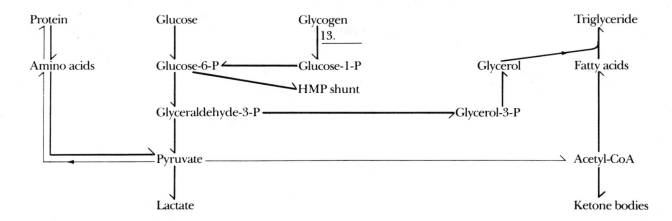

The insulin/glucagon ratio in the serum reflects the metabolic state of the person. This ratio peaks soon after commencing a meal at approximately 0.5 and then falls during fasting to reach very low levels (e.g., 0.05) within three days.

Answers

1. Insulin, somatomedins, and to a lesser extent, growth hormone, androgens, and thyroid hormones promote protein synthesis in the fed state.

2, 3. By inhibiting gluconeogenesis, insulin promotes glycolysis.

4, 5. Insulin and somatomedins promote the synthesis of glycogen and triglyceride.

6. Glucagon and epinephrine hasten glycogen breakdown in early starvation.

7. The main lipolytic hormones of early starvation are glucagon and epinephrine; growth hormone and thyroid hormones play a lesser role.

8, 9, 10, 11. In late starvation, glucocorticoids promote gluconeogenesis with attendant protein catabolism.

12. Glucagon and epinephrine are the main lipolytic hormones of late starvation. Growth hormone and thyroid hormones also promote lipolysis.

13. During anaerobic metabolism, epinephrine and glucagon mobilize glucose from glycogen.

References

Devlin, T. M. *Textbook of Biochemistry with Clinical Correlations* (3rd ed.). New York: Wiley-Liss, 1992. Pp. 576–603.

Mathews, C. K., and van Holde, K. E. *Biochemistry.* Redwood City, Calif.: Benjamin/Cummings, 1990. Pp. 779–811.

Murray, R. K., Granner, D. K., Mayes, P. A., and Rodwell, V. W. *Harper's Biochemistry* (22nd ed.). Norwalk, Conn.: Appleton & Lange, 1990. Pp. 261–266.

Stryer, L. *Biochemistry* (3rd ed.). New York: Freeman, 1988. Pp. 627–645.

Review Problems

This set of review problems highlights some of the key issues of medical biochemistry and will help you to identify those areas that require further study. It is not, however, an exhaustive test on all aspects of biochemistry.

Problem 1
The debate continues concerning the toxicity to humans of ammonia in high concentrations. Which of the following reactions does *not* dispose of ammonia?

    A. Carbamoyl-phosphate synthetase.
    B. Glutamine synthesis from glutamate.
    C. Aspartate synthesis from oxaloacetate.
    D. Glutamate synthesis from α-ketoglutarate.
    E. Arginine synthesis from argininosuccinate.

Problem 2
Carnitine deficiency in humans leads to muscle weakness and liver dysfunction partly because of the accumulation in muscle and liver of which one of the following substances?

    A. Glycogen.
    B. Odd-carbon fatty acids.
    C. Sphingolipids.
    D. Triglycerides.
    E. Dolichol phosphate.

Problem 3
From the statements given below, select the *false* statement about xanthine oxidase.

    A. Reduces O_2 to superoxide anion (O_2^-).
    B. Requires molybdenum.
    C. Catalyzes two successive reactions of purine catabolism.
    D. Requires iron.
    E. Inhibited by allopurinol.

Problem 4
By inhibiting the carboxylation of acetyl-CoA to yield malonyl-CoA, glucagon will:

A. Promote biosynthesis of fatty acids.
B. Inhibit ketogenesis.
C. Inhibit β-oxidation of fatty acids.
D. Promote ketone body formation.
E. Reduce gluconeogenesis.

Problems 5–7
Match the following methyl-donors to the methylation reactions given in Problems 5–7.

A. *S*-Adenosylmethionine
B. CH_2—THFA
C. Carboxybiotin
D. Tetrahydrobiopterin

5. dUMP $\longrightarrow$ dTMP (thymidylate synthetase reaction)
6. Norepinephrine $\longrightarrow$ epinephrine
7. Phosphatidylethanolamine $\longrightarrow$ phosphatidylcholine

Problem 8
A *noncompetitive* enzyme inhibitor will do all of the following *except:*

A. Decrease V_{max}.
B. Act independently of [S].
C. Decrease K_m.
D. Not attach to a substrate binding site.

Problem 9
The enzymes regulated by phosphorylation/dephosphorylation (e.g., covalent binding and later cleavage of phosphate to the enzyme) include all of the following enzymes *except:*

A. Glucose-6-phosphatase.
B. Glycogen synthase.
C. Pyruvate dehydrogenase.
D. Glycogen phosphorylase.

Problems 10–21
In the following problems, draw the chemical structure of the missing intermediate and name any vitamins and coenzymes required for each reaction.

10.

$$CH_3-\overset{\overset{\text{O}}{\|}}{C}-CH_2-COOH \longrightarrow \underline{\hspace{3cm}} + CO_2$$
Acetoacetate   Acetone

11.

$$\text{}^-\text{OOC}-\text{CH}_2-\underset{\underset{\text{OH}}{|}}{\overset{\overset{\text{COO}^-}{|}}{\text{C}}}-\text{CH}_2-\text{COO}^- \longrightarrow \underline{\qquad\qquad} + \text{H}_2\text{O}$$

Citrate   *Cis*-aconitate

12.

$$\text{CH}_3-\overset{\overset{\text{O}}{\|}}{\text{C}}-\text{S}-\text{CoA} + \text{HCO}_3^- + \text{ATP} \longrightarrow$$

Acetyl-CoA

$$\underline{\qquad\qquad} + \text{ADP} + \text{P}_\text{i}$$

Malonyl-CoA

13.

$$\text{}^-\text{OOC}-\text{CH}_2-\text{CH}_2-\overset{\overset{\text{O}}{\|}}{\text{C}}-\text{COO}^- + \text{CoA} + \text{NAD}^+ \longrightarrow$$

α-Ketoglutarate

$$\underline{\qquad\qquad} + \text{CO}_2 + \text{NADH} + \text{H}^+$$

Succinyl-CoA

14.

$$\underline{\qquad\qquad} + \text{ADP} \longrightarrow \text{}^-\text{OOC}-\overset{\overset{\text{O}}{\|}}{\text{C}}-\text{CH}_3 + \text{ATP}$$

PEP   Pyruvate

15.

$$\text{CO}_2 + \text{NH}_4^+ + \text{H}_2\text{O} + 2\text{ATP} \longrightarrow$$

$$\underline{\qquad\qquad} + 2\text{ADP} + \text{P}_\text{i} + 3\text{H}^+$$

Carbamoyl phosphate

16.

$$\text{}^-\text{OOC}-\overset{\overset{\text{O}}{\|}}{\text{C}}-\text{CH}_3 + \text{CO}_2 + \text{ATP} \longrightarrow \underline{\qquad\qquad} + \text{ADP} + \text{P}_\text{i}$$

Pyruvate   Oxaloacetate

17.

$$\underline{\qquad\qquad} + 2\text{NADPH} + \text{H}^+ \longrightarrow$$

HMG-CoA

$$\text{}^-\text{OOC}-\text{CH}_2-\underset{\underset{\text{CH}_3}{|}}{\overset{\overset{\text{OH}}{|}}{\text{C}}}-\text{CH}_2-\text{CH}_2\text{OH} + 2\text{NADP}^+ + \text{CoA}$$

Mevalonic acid

18.

$$\text{HOCH}_2-\underset{\underset{\text{Serine}}{|}}{\overset{\overset{\text{}^+\text{NH}_3}{|}}{\text{CH}}}-\text{COO}^- + \text{Coenzyme} \longrightarrow$$

$$\underline{\qquad\qquad} + \text{CH}_2-\text{Coenzyme} + \text{H}_2\text{O}$$

Glycine

19.

$$\text{}^-OOC-CH_2-CH_2-\overset{\overset{\displaystyle +NH_3}{|}}{CH}-COO^- + NH_4{}^+ + ATP \longrightarrow$$

Glutamate

$$\underline{\hspace{3cm}} + ADP + P_i + H^+$$
Glutamine

20.

$$\text{}^-OOC-CH_2-\overset{\overset{\displaystyle O}{\|}}{C}-COO^- + \text{Glutamate} \longrightarrow$$

Oxaloacetate

$$\underline{\hspace{3cm}} + \alpha\text{-Ketoglutarate}$$
Aspartate

21.

$$\text{R}-\overset{\overset{\displaystyle }{|}}{\underset{\underset{\displaystyle OH}{|}}{CH}}-CH_2-\overset{\overset{\displaystyle O}{\|}}{C}-S-CoA + NAD^+ \longrightarrow$$

β-Hydroxyacyl-CoA

$$\underline{\hspace{3cm}} + NADH + H^+$$
β-Ketoacyl-CoA

Problem 22

Gluconeogenesis can proceed from all of the following *except:*

    A. Oxaloacetate.
    B. Pyruvate.
    C. Propionyl-CoA (from odd-carbon fatty acid breakdown).
    D. Hydroxybutyrate.
    E. Citrate.

Problem 23

NADPH, instead of NADH, is required in the biosynthesis of all of the following *except:*

    A. Testosterone.
    B. Fatty acids.
    C. Cholesterol.
    D. DNA.
    E. Phosphatidylcholine.

Problem 24

The four enzymes below require pyridoxal phosphate as a coenzyme. For which enzyme does this requirement seem *unexpected* in view of the usual role of pyridoxal phosphate?

    A. Glutamate-oxaloacetate transaminase (GOT).
    B. Alanine 2-oxoglutarate aminotransferase.
    C. Glycogen phosphorylase.
    D. Serine dehydratase.

Problem 25
Tetrahydrobiopterin is used as a coenzyme in the metabolism of:

  A. Folic acid.
  B. Phenylalanine.
  C. Biotin.
  D. Asparagine.
  E. Valine.

Problems 26–30
Match the mechanisms of action or toxicity in humans described below with the agent/enzyme pairs given in Problems 26–30.

  A. Prevents substrate binding to a non-enzyme protein
  B. Competitive enzyme inhibitor
  C. Non-competitive enzyme inhibitor
  D. Enzyme denaturation

26. Lead/many enzymes
27. Carbon monoxide/hemoglobin
28. Hydroxycitrate/citrate lyase
29. KOH/many enzymes
30. Carbidopa/dopa decarboxylase

Problem 31
Choose the enzyme involved in gluconeogenesis but *not* in glycolysis.

  A. Hexokinase
  B. Glucose-phosphate isomerase
  C. Enolase
  D. Lactate dehydrogenase
  E. PEP carboxykinase

Problem 32
Tyrosine kinase:

  A. Converts tyrosine to phenylalanine.
  B. Is stimulated by catabolic hormones.
  C. Activates calmodulin.
  D. Explains the mechanism of action for thyroid hormones.
  E. Phosphorylates tyrosine in cytosolic proteins.

Problem 33
The polymerase chain reaction involves all of the following steps except:

  A. The DNA fragment to be amplified (call it X) is inserted into a plasmid vector whose DNA sequence is known.
  B. The plasmid vector is heated to 90°C to separate the two DNA strands and then cooled, adding oligonucleotides complementary to the known vector DNA strands.

C. The oligonucleotides bind to the plasmid vector DNA on either side of the X fragment.
D. These oligonucleotides act as primers for DNA polymerase, which copies both DNA strands of X.
E. DNA X and the plasmid vector DNA are copied. The plasmid vector DNA is separated by electrophoresis and DNA X is isolated.

Problem 34

To treat genetic diseases by gene replacement, recombinant DNA containing the desired gene can be introduced in vitro into host cells. This process is known as **transfection.** Methods for transfection of human cells include all of the following except:

A. Exposure to calcium phosphate, which promotes DNA uptake by host cells.
B. Use of bacterial vectors.
C. Retroviruses.
D. Injection of foreign DNA into host cells.
E. Certain DNA viruses.

Problems 35–36

Following is a list of steps that may or may not be involved in DNA replication.

A. DNA polymerase adds NTPs in a 5′ to 3′ direction.
B. Primase creates an RNA primer.
C. DNA ligase joins the DNA fragments.
D. Unwinding proteins open the DNA double helix.
E. DNA polymerase fills the gap between Okazaki fragments and removes the RNA primer.
F. DNA polymerase removes the RNA primer.

35. Place the steps of DNA replication for the *leading* strand in proper sequence.
36. Place the steps of DNA replication for the *lagging* strand in proper sequence.

Problem 37

Choose the reaction that consumes two high-energy phosphate bonds due to conversion of ATP to AMP + PP_i.

A. Amino acid + tRNA $\longrightarrow$ aminoacyl-tRNA
B. Glucose-1-P + UTP $\longrightarrow$ UDP-glucose + PP_i
C. Oxaloacetate $\longrightarrow$ PEP + CO_2
D. Pyruvate + CO_2 $\longrightarrow$ oxaloacetate
E. Fructose-6-P $\longrightarrow$ fructose-1,6-diP

Problem 38
Choose the incorrect statement about G proteins:

A. Glucagon or epinephrine binding to receptors converts G proteins to their active forms.
B. The name G proteins stands for guanyl-nucleotide-binding proteins.
C. The active GTP form stimulates adenyl cyclase, leading to cyclic AMP production.
D. One subunit contains a GTPase that hydrolyzes the active form back to the inactive one.
E. They regulate tyrosine kinase.

Problem 39
Which vitamin *cannot* serve as an antioxidant?

A. Vitamin A.
B. Ascorbic acid.
C. Vitamin E.
D. Vitamin K.

Problem 40
Which of the following steps is *not* a function of the pentose phosphate pathway?

A. Interconverts hexoses and pentoses.
B. Produces NADPH.
C. Supplies ribose-5-P.
D. Converts galactose to glucose.

Problem 41
Which of the following pathways occurs in part in the mitochondria and in part in the cytoplasm?

A. TCA cycle starting with acetyl-CoA and OAA.
B. Urea cycle.
C. Gluconeogenesis starting from OAA in the cytoplasm.
D. Glycolysis to lactate.
E. Oxidative phosphorylation.

Problem 42
People with diabetes mellitus are prone to develop cataracts because their elevated serum glucose concentrations:

A. Increase glycogen synthesis within the lens.
B. Lead to harmful levels of glucose-6-phosphate.
C. Inhibit gluconeogenesis.
D. Allow aldose reductase to reduce glucose to sorbitol.
E. Glycosylate hemoglobin.

Problem 43
Treatment of acute renal failure might include all of the following except:

A. Low-protein diet.
B. High-carbohydrate diet.
C. α-Keto acid analogs to the essential amino acids combined with a protein-free diet.
D. Low-fat diet.

Problem 44
Select the correct response with respect to tRNA:

A. The amino acid carried by each tRNA is important in selective binding of tRNA to codons.
B. Formation of aminoacyl-tRNA consumes GTP.
C. The 5' and 3' ends of tRNA are on opposite arms of the cloverleaf.
D. Each anticodon binds to one and only one codon.
E. Ribonuclease P helps convert the primary transcripts for tRNA into active tRNA.

Problem 45
Which mechanism is used to edit translation?

A. During the delay between the arrival of an aminoacyl-tRNA at the A site and peptide bond formation, incorrect aminoacyl-tRNA molecules detach from mRNA.
B. Peptidyl transferase detects incorrect aminoacyl-tRNA molecules and deletes them.
C. Elongation factors edit the growing peptide chain.
D. Releasing factors bind to incorrect aminoacyl-tRNA molecules and promote their removal from the peptide chain.
E. DNA polymerase edits during translation.

Problem 46
Place the steps in eukaryotic translation into the proper sequence.

A. Met-tRNA and mRNA bind to the ribosomes.
B. Uncharged tRNA leaves the P site, and the peptidyl-tRNA at the A site moves to the now-vacant P site.
C. Elongation factors bring aminoacyl-tRNA to bind to the A site.
D. Initiation factors bind to the ribosomes.
E. Peptidyl transferase forges a peptide bond.
F. Releasing factors bind to a stop signal, causing the peptidyl transferase to hydrolyze the bond joining the polypeptide at the A site to the tRNA.
G. The ribosome advances three bases to read the next codon.

Problem 47
The lactose operon:

A. Includes structural genes, control sites, and regulator genes.
B. Is regulated by lactose alone.
C. Has attenuation sites.
D. Is controlled in part by cyclic AMP.
E. Is normally switched on and must be repressed to be switched off.

Problem 48
The conversion of succinate to fumarate in the citric acid cycle resembles that of acyl-CoA to enoyl-CoA in fatty acid oxidation because both reactions involve:

A. Decarboxylation.
B. Oxidation by NADH.
C. Phosphorylation.
D. Oxidation by FAD.
E. Transamination.

Problems 49–52
Match the organelles below to the functions in Problems 49–52.

A. Endoplasmic reticulum
B. Golgi apparatus
C. Lysosomes
D. Ribosomes

49. Terminal glycosylation of glycoproteins.
50. Main function is degradation.
51. Sorts proteins and packages certain proteins in vesicles.
52. Core glycosylation of glycoproteins.

Problem 53
Which of the following nonproteins can act as an enzyme?

A. DNA.
B. Phospholipids.
C. Glycolipids.
D. RNA.
E. Magnesium.

Problem 54
Choose the incorrect statement about uracil in DNA.

A. Formed from cytosine.
B. Uracil-DNA glycosidase removes uracil from DNA.
C. After uracil is removed from DNA, DNA polymerase replaces it with adenosine.
D. DNA ligase then seals the repaired strand.

Problems 55–57

Match the following structures to the descriptions in Problems 55–57.

    A. Introns
    B. Exons
    C. Splicesomes
    D. Operons

55. A group of related bacterial genes controlled by the same signals.
56. Ribonucleoprotein complexes that hold exons together during one phase of transcription.
57. Untranslated portions of eukaryotic genes.

Problem 58

Fat is a superb energy store. Which of the following does *not* explain the need to have glycogen stores?

    A. Glycogen can provide fuel more quickly to exercising muscle than fat.
    B. Fat stores more energy per unit weight than glycogen.
    C. Glycogen can be utilized in the absence of oxygen.
    D. No component of fat can be converted to glucose.
    E. Glycogen can be converted to fatty acids.

Problem 59

A child has an enlarged liver due to excess glycogen stores. The glycogen is of normal structure. There is no rise in serum glucose concentration after the oral administration of glycogen. These observations suggest that this disorder results from the absence of:

    A. Fructose-1-P aldolase.
    B. Fructokinase.
    C. Glucokinase.
    D. Amylo-1,6-glucosidase.
    E. Glucose-6-phosphatase.

ANSWERS

1. E.
2. D.
3. A. (H_2O_2, not O_2^-)
4. D.
5. B.
6. A.
7. A.
8. C.
9. A.

10. 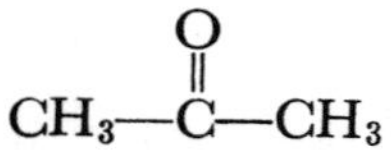
$$CH_3-\overset{\overset{\displaystyle O}{\|}}{C}-CH_3$$

11.
$$^-OOC-CH_2-\overset{\overset{\displaystyle COO^-}{|}}{C}=CH-COO^-$$

12.
$$CH_3-CH_2-\overset{\overset{\displaystyle O}{\|}}{C}-S-CoA.$$ Requires biotin, pantothenic acid (in CoA).

13.
$$^-OOC-CH_2-CH_2-\overset{\overset{\displaystyle O}{\|}}{C}-S-CoA.$$ Requires pantothenic acid, niacin (in NAD^+), thiamine, lipoic acid (a coenzyme, not a vitamin), and riboflavin (in FAD^+).

14.
$$^-OOC-\overset{\overset{\displaystyle O-\text{\textcircled{P}}}{|}}{C}=CH_2$$

15.
$$H_2N-\overset{\overset{\displaystyle O}{\|}}{C}-O-\text{\textcircled{P}}$$

16.
$$^-OOC-\overset{\overset{\displaystyle O}{\|}}{C}-CH_2-COO^-.$$ Requires biotin.

17.
$$^-OOC-CH_2-\overset{\overset{\displaystyle OH}{|}}{\underset{\underset{\displaystyle CH_3}{|}}{C}}-CH_2-\overset{\overset{\displaystyle O}{\|}}{C}-S-CoA.$$ Requires pantothenic acid, niacin (in NADPH). (This is the rate-controlling step of cholesterol biosynthesis)

18. $\overset{\displaystyle ^+NH_3}{\underset{|}{}}$
CH_2-COO^-. Folic acid (in THFA).

19.
$$O=\overset{\overset{\displaystyle NH_2}{|}}{C}-CH_2-CH_2-\overset{\overset{\displaystyle ^+NH_3}{|}}{CH}-COO^-$$

20.
$$^-OOC-CH_2-\overset{\overset{\displaystyle ^+NH_3}{|}}{CH}-COO^-.$$ Requires pyridoxine (in pyridoxal phosphate).

21.
$$R-\underset{\underset{\displaystyle O}{\|}}{C}-CH_2-\overset{\overset{\displaystyle O}{\|}}{C}-S-CoA.$$ Requires niacin.

22. D.
23. E.
24. C.
25. B.
26. C.
27. A.
28. B. (This investigational drug inhibits conversion of citrate to ox-aloacetate plus acetate. Thus, animals given a low-fat, high-calorie diet along with hydroxycitrate will not become obese because they will have insufficient acetate for fatty acid biosynthesis. Alas, another panacea for obesity!)
29. D.
30. B. (Carbidopa reduces the dosage of L-dopa needed to treat Parkinson's disease.)
31. E.
32. E.
33. E. The plasmid vector DNA is not copied.
34. B. Bacteria cannot transfect human cells.
35. D, B, A, F.
36. D, B, A, E, C.
37. A.
38. E.
39. D.
40. D.
41. B.
42. D. Once formed, sorbitol cannot easily leave the lens. Sorbitol dehydrogenase cannot convert enough of this sorbitol to fructose.
43. D. High-carbohydrate diet will reduce gluconeogenesis, thereby slowing protein catabolism. α-Keto acid analogs to the essential amino acids will combine with ammonia to form essential amino acids. Thus, protein synthesis can be maintained without adding additional nitrogen to a person unable to excrete nitrogen.
44. E. The amino acid on tRNA has no role in selective binding to codons. Aminoacyl-tRNA formation consumes ATP, not GTP. The 5′ and 3′ ends of tRNA are on the same arm of the cloverleaf. Some anticodons bind to more than one codon.
45. A.
46. D, A, C, E, B, G, F.
47. D. The regulator gene lies outside operons. Allolactose, a metabolite of lactose, is the primary inducer for the lactose operon. The tryptophan operon has attenuation sites but the lactose operon does not. Cyclic AMP binding to CAP (catabolite gene-activating protein) plays a part in regulating the lactose operon. The lactose operon is normally switched off.
48. D.
49. B.
50. C.
51. B.

52. A.
53. D. Certain RNA has RNA polymerase activity.
54. C. DNA polymerase replaces uracil with cytosine.
55. D.
56. C.
57. A.
58. D. The glycerol backbone of triacylglycerols can be converted to glucose. Fat stores more energy per unit weight than glycogen but this fact would in isolation argue against the need for glycogen stores. Glycogen can be converted to fatty acids but this would not explain the need to store glycogen.
59. E.

Index